The Food Question:

Profits Versus People?

The Food Question:

Profits Versus People?

edited by
Henry Bernstein
Ben Crow
Maureen Mackintosh and
Charlotte Martin

Routledge
Taylor & Francis Group
LONDON AND NEW YORK

First published 1990 by Earthscan Publications Ltd

2 Park Square, Milton Park, Abingdon, Oxon OX14 4RN
711 Third Avenue, New York, NY 10017, USA

Routledge is an imprint of the Taylor & Francis Group, an informa business

First issued in hardback 2016

British Library Cataloguing in Publication Data
The food question.
1. Developing countries. Food supply
I. Bernstein, Henry
338. 1'9' 1724
ISBN: 978-1-85383-063-1 (pbk)
ISBN: 978-1-138-16603-5 (hbk)

Production by David Williams Associates (01–521 4130)
The Guernsey Press Co. Ltd., Guernsey, Channel Islands.

Contents

Contributors vii

1 Introduction 1
The editors

2 The Origins of Third World Food Dependence 13
Harriet Friedmann

3 Moving the Lever: a New Food Aid Imperialism? 32
Ben Crow

4 Abstract Markets and Real Needs 43
Maureen Mackintosh

5 Food Crisis and Gender Conflict in the African Countryside 54
Ann Whitehead

6 Taking the Part of Peasants? 69
Henry Bernstein

7 Some Economic and Political Consequences of the Green Revolution in India 80
Utsa Patnaik

8 Another Awkward Class: Merchants and Agrarian Change in India 91
Barbara Harriss

9 Agrarian Crisis and Political Crisis in Mexico 104
Roger Bartra

10 "Structural Adjustment", Agribusiness and Rural Women in Tanzania 111
Marjorie Mbilinyi

11 Hunger and Women's Survival in a Bangladesh Slum 125
Jane Pryer

12 Poverty, Purdah and Women's Survival Strategies in Rural Bangladesh 134
Naila Kabeer

13 Peasants Under Contract: Agro-food Complexes in the Third World 149
Michael Watts

14 Biotechnology and Agricultural Development in the Third World 163
Frederick H. Buttel

15 Urban Consumptionism as a Route to Rural Renewal 181
Robin Jenkins

16 Further Resources on Food 196
Charlotte Martin

Index 204

Contributors

Roger Bartra is Professor at the Instituto de Investigaciones Sociales, Ciudad Universitaria, Mexico City, and author of many books and articles with a major influence on agrarian debates in Latin America.

Henry Bernstein teaches at the Institute for Development Policy and Management, University of Manchester, and is a member of the Development Policy and Practice research group (DPP) at the Open University, editor and co-editor of three books and, with T.J. Byres, of the *Journal of Peasant Studies*.

Frederick H. Buttel is Professor of Rural Sociology and Faculty Associate in the Program on Science, Technology and Society, Cornell University, with interests in agrarian change, biotechnology and environmental sociology; also author and editor of several books.

Ben Crow is Lecturer in Development Studies and co-chair of DPP at the Open University; interests include a wide range of agrarian and environmental issues; currently researching rice markets in Bangladesh. He is co-editor of *Survival and Change in the Third World* and co-author of the *Third World Atlas*.

Harriet Friedmann teaches Sociology at the University of Toronto and is the author of a forthcoming book on the *Political Economy of Food*.

Barbara Harriss is Lecturer in Agricultural Economics at Oxford University and an associate member of DPP; has spent many years' research on the economics and politics of grain trading in South Asia and West Africa and is author of several books in this area.

Robin Jenkins was Food Policy Adviser at the Greater London Council until its abolition, and is now Catering Manager for schools, social services and staff canteens in Hackney, Britain's poorest borough; has been an academic and farmer; author of books on imperialism (*Exploitation*, 1970) and peasant farming (*The Road to Alto*, 1979).

Naila Kabeer is a Research Fellow at the Institute of Development Studies University of Sussex. She specializes in research and training on gender and development issues.

Maureen Mackintosh teaches Economics at Kingston Polytechnic, UK, and is a member of DPP, with research experience in Mozambique and West Africa; author of *Gender, Class and Rural Transition: Agribusiness and the Food Crisis in Senegal*.

Charlotte Martin is a school teacher in London and a part-time tutor for the Open University's Third World Studies course; interests include educational materials in Third World development for use in schools.

Marjorie Mbilinyi is Professor of Development Studies, University of Dar es Salaam, and has researched and written extensively on gender in Tanzania in relation to education, rural society and agrarian capitalism.

Utsa Patnaik is Professor of Economics at Jawarhalal Nehru University, New Delhi; author of *Peasant Class Differentiation* (1987), and a leading contributor to agrarian debates in India.

Jane Pryer is a social nutritionist and Research Associate at the London School of Hygiene and Tropical Medicine, and co-author of *Cities of Hunger* (1988).

Michael Watts teaches Geography and Development Studies at the University of California, Berkeley; author of *Silent*

Violence (1983) on food, famine and peasantry in colonial Nigeria; currently researching agricultural restructuring in West Africa and California.

Ann Whitehead teaches Social Anthropology and Women's Studies at the University of Sussex, and is on the editorial collective of *Feminist Review*; has a longstanding research interest in socio-economic change and gender relations in rural Africa.

1 Introduction

The editors

Recent years have seen an increasing awareness of the urgency of food problems, at their most dramatic in the contrast between wasteful overproduction and overconsumption (by some) of food in the developed capitalist countries, and the continuing, in some cases worsening, hunger of many millions of people in the Third World.

This awareness and the concern it generates have been stimulated by a critical campaigning literature in which books like Frances Moore Lappé and Joseph Collins's *Food First* and Susan George's *How the Other Half Dies* are notable landmarks. These authors and others have done much to express and give focus to the growing sense of the inequalities and injustices in the world economic system, and the brutalities resulting from them including the stark facts on hunger.

At the same time we believe that the anger felt by many needs to be matched by an adequate analysis of the sometimes complex reasons for the obscene connections between abundance and scarcity, if radical action for change is to be effective. For those who are concerned and angry, and want to work for change, there is a need to combine emotional commitment with a careful investigation of moral judgements (such as that small farmers are "good" while multinational companies are "bad") and frequently cited "facts". It is certainly important to get at the facts, and there are vested interests that try to prevent this (see Jenkins's comments on the British Nutrition Foundation in Chapter 15 of this volume), but however powerful the facts appear they do not "speak for themselves" (or only do so for those already converted). Facts only become *evidence* necessary to win arguments when they are combined with *analysis* and meet its intellectual demands and responsibilities.

The aim of this collection, then, is to contribute some tools of analysis, and to illustrate their applications. We hope to make the results of specialized research more accessible to all those concerned with the issues, so that they are better equipped to confront and assess for themselves the theoretical, political and practical challenges involved.

This aim is also a guiding principle of the research group on Development Policy and Practice (DPP) at Britain's Open University with which we are associated. DPP is engaged in research on aspects of the food question in South Asia, Central and Southern Africa, and Central America, in collaboration with people and institutions in those regions. This collection, therefore, draws on the research and political involvement of colleagues and friends in the Third World, and also in Britain and in North America.

The contributors write from a broadly socialist position. Collectively we have tried both to draw on the historic strengths of socialist analysis concerning social classes and social relations of production and power, and to recognize the need for the contemporary and future agenda of socialism to engage just as seriously with issues of gender, of exchange, of distribution and consumption (as well as production), and of the environment. Hence this book is a contribution to the "political economy" of the food question.

A political economy of food

What do we mean by a political economy of food? At the simplest level, all the chapters in this book examine the interrelations between politics and economics: between political ideas and actions and economic ideas and processes. To do this, they all examine aspects of the social organization, or social relations, of the production, exchange, distribution and consumption of food. And they focus on the contradictions and conflicts which structure these social relations and which drive economic change: class divisions between those who benefit from the work of others and those who are exploited; conflicts between men and women over production and access to food; conflict between governments and between governments and people. Finally, all are concerned with actions and responses: examining ways in which people struggle against their situation and discussing potential responses to the situations analysed.

All chapters explore the political economy of capitalism rather than elaborating alternatives. This is because capitalism has created the world we inhabit and also because it is necessary to understand the complexities and contradictions of "actually existing capitalism" (as opposed to simplistic models of capitalism, whether of an ideologically positive or negative kind) in order to understand the prospects and tasks of radical change.

The food question is a very large one. While we can only cover a fraction of the relevant issues, the book does contain analyses of many different levels of the political economy of food. The chapters examine problems of access to food in city and countryside; changes in production relations in food farming, especially the connections between small peasant farmers and large multinational firms and agencies; how markets work and their effects on local, national and international levels; and the "development" strategies, in relation to food, of the large agencies such as the World Bank and the IMF. The central concern is who gets to eat what – and, especially, why?

We said earlier that socialists have been forced to broaden their agenda from a concentration on class and production to include other social relations within which these are embedded. This is strikingly illustrated in this collection by the emphasis in a number of chapters on analysing markets and on examining relations between men and women.

Socialists have in the past been rather uninterested in the

detail of how markets function. They have tended, following Marx, to see markets as "surface" phenomena, disguising the real underlying processes of exploitation. But, while it is certainly the case that a restriction of economic analysis to markets alone does allow the apparent freedom and relative equality of exchange in markets to be used to conceal lack of freedom and unequal power within production, nevertheless, markets are crucial to the maintenance of exploitative and unequal production relations. Markets channel the results of production, perpetuating the control by some of the activities and consumption of others. They are a crucial link in the system which determines who eats, how much and when. Markets, furthermore, are diverse. Markets for huge sums of capital are quite different – in structure and institutions, in the power participants wield, in their implications – from local markets. One of the aims of this book is to examine the variety of markets and their implications for those who participate in them, and for those who wish to change them. The emphasis is on markets for food itself, but markets for labour, land and inputs to food production are also important.

Markets also have diverse impact on production; the creation of markets, which is one of the aims of many international agencies, influences production and consumption in varied ways in different regions, for different classes, and for men as opposed to women. Several chapters examine this issue. Perhaps the key point to emerge is that markets must not be conceptualized as the manifestation of impersonal laws of demand and supply or of a "hidden hand", but are social institutions and processes that people enter from very different positions in structures of class, gender and power. This in turn affects what they get from markets to secure their livelihoods, to satisfy – or fail to satisfy – their basic needs. There exist no "free" markets without different types of state "intervention", formal or informal, hence no satisfactory economics of markets without a politics of markets.

The second area is that of gender relations. Many of the

images of food production, or food markets, or hunger, which we have been faced with in recent years have been images of women: women hoeing barren land; women market traders in Africa; women looking after starving children. People have begun to talk about the feminization of poverty, and of women being "left behind" by the processes of development. Yet oddly, while women now form so important a part of our *images* of the food question, so much *writing* about food and hunger separates the role of women from the analysis of other aspects of the problem, tacking them on at the end as victims or as burdened workers. There is much good writing about women and development, but books *not* about "women and . . ." still tend to treat women as a separate issue.

We have tried, not wholly successfully, to avoid this division. The development of food production and of access to food today cannot be understood unless we add to the "kit" of ideas we use to analyse the world, the concept of gender. Gender is different from biological sex. It expresses and investigates the *relations* between men and women, the constructed social positions and the circumscribed range of activities of men and women which are crucial elements in understanding how society and economy work. It has embedded in it the inequalities between men and women, and the conflicts between them, which are constantly recreated within an exploitative system, and which form a central dynamic in the creation and maintenance of poverty.

Old arguments therefore about whether "class" or "gender" is more important are misleading. Classes are "gendered". The experience of men and women of one class, though intertwined, is different: "women's work" and "men's work", recognizable categories in all societies, are quite distinct. Also, the genders are divided internally by class, in terms again of their experience of the world: what it is to be male or female is deeply determined by the class within which one lives.

If class and gender are so deeply intertwined then, in our unequal and exploitative societies, the feminization of poverty is a largely inevitable process, though the *extent* of the poverty and lack of access of the women of the poorest classes can still come as a severe shock. And as inequality grows, through the deepening divisions being enforced on people in the Third World in the wake of recession in the West, so women – and therefore children – increasingly take the brunt: hence the images on the TV screens. Many chapters in this book examine these complexities, focusing on sexual divisions of labour and on women's access to food.

One of the implications of the feminization of poverty is that women have often been in the forefront of struggles over food: over access to food; over the quality of food; over the conditions of its production. Several chapters examine women's efforts to feed their households: "survival strategies" in conflict-ridden circumstances. They also describe active resistance to changing terms of food production and the scope for collective response.

This theme of response to pressure is the final issue which we want to draw out of the book as a whole. A true political economy examines not merely what is happening, but also what is being done, how people respond and resist, what *can* be done. All the contributors pick up this theme. Some analyse particular forms of resistance; others examine the political implications of changes in the rural economy; others point out the implications for political strategies in the Third World or in the West of developments in international and national markets.

It is in the discussion of struggle that the divisions and oppressions of labour by capital emerge most strongly in this book. While the view of capitalism as a mode of production based on private capital ownership and labour "free" to sell its services for a wage remains a central theme of political economy, this book shows the extent to which this has to be tempered by a recognition of the acute limitations set by capitalism on the freedoms of labour. The complexity of the experiences of labour under capitalism is illustrated in this book by the material on class and gender, the analysis of household and contract labour, and the discussion of the casualization of wage work. Resistance to the conditions of food production and other agricultural work is structured by the divisions which capitalism seeks to impose on those who work: divisions by gender, by urban and rural location, by nationality and by position in the organization of production. Several of the chapters address the question of how to find a basis for unity across these divisions.

We hope then that this collection will be a source book of ideas and concepts on the question of food, and that as such it will be useful to people who are seeking to be active on the politics of food.

The essays

Harriet Friedmann's contribution sets the scene both historically and conceptually with a broad and incisive survey of the post-war international food order and its contradictions. Its central focus is food (especially wheat) production in the United States and how it combined with both domestic farm policy and foreign policy, including food aid, to determine the patterns of world markets in wheat until the early 1970s, when this international food order started to disintegrate. The effects of American food surpluses, foreign policy and food aid for markets and production in many Third World countries, are analysed within a perspective which seeks to show how "food policies are an aspect of class politics, even though they work through international politics".

Ben Crow provides a detailed case study of American food aid to Bangladesh, also highlighting more general issues of the uses of Western aid to try to impose comprehensive policy "reform" on Third World countries. The conditionality exerted to push economic "development" along lines of market liberalization is expressed memorably in the words of a USAID official quoted: to get "more policy bang per buck".

Maureen Mackintosh's essay interrogates the question of markets with special reference to food and whether human needs are satisfied or not. She takes issue with both the ideological advocacy of "the" market as a solution to problems, and the problematic neglect by socialists of key issues about how markets work (reflecting also a simplistic dichotomy between markets and planning). Her contribution serves as a general overview to which other chapters dealing with more specific markets and market issues (e.g. by Friedmann, Harriss and Jenkins) can be related.

Ann Whitehead criticizes an over-simplification which has become common currency in discussions about women in the African countryside: the view that the food crisis in Africa is centrally the result of neglect in some sense of women's food farming, by both African men and international agencies. Instead, she argues that impoverishment and the struggle for food are faced by both men and women – they are a problem for all members of African rural households. However, men and women face the crisis differently, and in ways which cause conflict between them. She distinguishes the impact of markets (commoditization) for farm products and for land, the influence of gender relations in farming and marriage, and the misogyny of external agencies, in the emergence of general and of female impoverishment, and argues that we must not allow gender conflicts which have developed to be used to obscure the economic forces of capitalism which restructure and exacerbate rural impoverishment.

In his essay Henry Bernstein acknowledges the force of populist and romantic views of the peasantry as one kind of anti-capitalist ideology and critique of the effects of capitalism on agriculture and food production. He questions, however, the effectiveness and political direction of populist analysis, suggesting that the conditions of peasant existence in capitalism are structured in ways that differentiate peasants by relations of class and gender. It is not enough to advocate taking the part of the peasants but is necessary to ask, "Which peasants?"

This has key implications for different strategies to solve the food question, at the same time confronting contradictions in the countryside and between countryside and town that are generated by capitalism.

Two pieces on India concentrate respectively more on patterns and trends of food production and their consequences, and on how food markets work, though both also eloquently illustrate patterns of spatial and regional differentiation and inequality characteristic of capitalism. Utsa Patnaik provides a cogent review of some economic and political consequences of the Green Revolution that casts a necessarily critical light on the now conventional wisdom that it has solved the food problems of India, or more precisely of India's poor. Her analysis suggests links between class and regional differentiation associated with the Green Revolution, illustrated in particular by the formation of an agrarian-based bourgeoisie and its role in separatist politics in the Punjab.

Barbara Harriss's piece on the organization of grain markets in India highlights the power of merchants (in particular large merchants), and its significance for agrarian change and economic development more generally. The characteristics of this "awkward class" confound many of the expectations of both advocates and some critics of capitalist development. The accumulation of profits and capital by dominant classes of merchants is not necessarily channelled towards productive reinvestment, any more than the kinds of market power she analyses help the great majority of producers and consumers of food.

Roger Bartra provides an interpretation of Mexico that summarizes a great deal of that country's rich and complex agrarian history. He argues that a historic land reform provided the basis of a certain kind of state formation and political stability which enabled capitalist investment in agriculture to take place at the expense of peasant production. The limitations of the land reform and contradictions of Mexico's particular pattern of capitalist agricultural development generated an agrarian and food crisis that in turn is a leading factor in the national political crisis.

The next piece by Marjorie Mbilinyi shows how the World Bank's efforts to restructure the Tanzanian economy rely on intensifying women's work in all kinds of farming, for home consumption and for the market, and on pushing women into casualized agricultural labouring. She documents the implications of this for women, and the nature and scale of their resistance to this pressure. As also suggested by Ann Whitehead, an existing sexual division of labour is being transformed under pressure into an intolerable double burden which women seek to escape, not least by migration. There is a danger that poorly thought-out "women in development" programmes, if they do not start from support for women's efforts to resist and to restructure their own situation, may in practice reinforce the efforts of other agencies to retain women as cheap casual labour on land no longer theirs.

Two further chapters in the very different context of Bangladesh also see women's social and economic subordination as central to an understanding of hunger and poverty. Jane Pryer describes women's struggle for access to food in a slum society detached from the land, measuring the scale of deprivation and struggle for survival generated by the particularly severe restrictions on the economic activities of women. Naila Kabeer uncovers the "survival strategies" forced upon rural women, in the face of class and patriarchal oppression, arguing that failure to examine the complex processes of female adaptation and resistance forced by the struggle to survive is one reason for the extraordinarily divergent public assessments of the extent and development of poverty and inequality in Bangladesh. Both of these chapters combine an examination of the context and the content of our images of female impoverishment, as well as giving life to the resistance and strength of the women those images refer to: women not as victims, but actively engaged in struggling to change their situation. Naila Kabeer ends on the same point as Marjorie

Mbilinyi: external "aid" has to take as its starting-point support for these struggles, and especially for the attempts of women to join together to pursue collective solutions.

The chapters by Michael Watts and Fred Buttel bring us back to the dynamics of international capitalism, adding important dimensions that complement Harriet Friedmann's analysis. Michael Watts looks at contract farming in Third World countries and how it is engineered by agribusiness multinationals, typically in alliance with "host"-country governments (and often facilitated by foreign aid). Watts not only documents and illustrates this as a major contemporary trend, but also suggests how it reflects important shifts in the ways that capital seeks to organize production and markets and to restructure international divisions of labour, in particular by creating new forms of bondage of labour to the production process. The observations about labour in contract farming provide a further illustration of our earlier point about the analysis of "actually existing capitalism", and the restrictions on the "freedom" of labour.

Fred Buttel focuses on biotechnology, the force for a new technical revolution in agriculture globally, the prospects and problems of which require a full understanding of how it is promoted and where it fits in existing patterns of capitalism. The principal point is that new biotechnologies are the "property" of agribusiness multinationals, and that their *technical* potential for increasing the quantity and quality of yields and of solving problems of food production is circumscribed by the *economic* interests of multinational capital in controlling markets and making profits. Buttel's international political economy of biotechnology also employs an important differentiation of Third World countries by their economic size and technical capacity, which further adds to ways of differentiating the "Third World" documented and analysed in other chapters in this book.

The final article by Robin Jenkins combines the results of theoretical, historical and political analysis with practical experience of food policies in Britain. He first sketches farming in Britain as a sector of particular capitalist interests, of exorbitant resource use in terms of energy and social costs, and of strong alliances with agribusiness, the food industry and the state. He then suggests how consumer power can be mobilized against this prevailing system to exert pressure for environmentally and socially more progressive ways of producing, processing and retailing food that would also have positive "knock-on" effects for agriculture and food production in Third World countries. The essay by Jenkins thus returns us squarely to the agenda of food politics with which the chapter by Harriet Friedmann concluded, posing again the question of political responses in the better-fed countries to the international problems of food and hunger.

2 The Origins of Third World Food Dependence

Harriet Friedmann

After 200 years of industrialization, the politics of food and agriculture remain critical to every region of the world economy and to the world economy as a whole. The politics of food reflect tensions between nations, between international financial agencies and populations who depend on food subsidies, between strategies of accumulation and self-determination. The reason, simply, is that food politics are an aspect of *class* politics, even though they work through *international* politics.[1]

The international food regime

There was a stable international food regime between roughly 1947 and 1972. It created a new set of relations between farmers and consumers, and a new set of trade relations among countries. The root of the regime was American *domestic* farm policy during a period of United States hegemony. The American farm programme supported commodities by in effect buying crops from farmers when their market prices were below the one targeted by the government (this took the form of loans at the beginning of the season to be repaid in grain at the set price or, if that were lower than the market price, in money). This created huge surpluses held by the government.

These surplus stocks had two effects on international wheat markets. First, they depressed prices and created problems for all other food producers in the world, enabling American farmers to displace others in, for example, Canada, Australia and Argentina, in export markets. But this depended on the second effect. The American government *created* new export markets in the Third World, in societies which were predominantly agrarian only a few decades ago – and in some cases still are. To understand the food import dependence of much of the Third World we must go back to the 1950s, which was the beginning of a period of rapid proletarianization, urbanization and changes in diet – mainly to American wheat.

The US state restructured international trade through the mechanism of *food aid*. In one way American aid was a type of dumping. But food aid was a *relationship* in which Third World states participated. They welcomed food aid as a way to support specific political projects of capitalist development which favoured cheap food (and low wages) over alternate capitalist or socialist projects focused on national food production. The counterpart to import dependence of the Third World is export dependence of First World farmers who, despite increasing government expenditures, are still going bankrupt under the weight of surplus stocks. The class content of the international food regime includes a deep (and deeply destructive) relationship between the urban poor of the Third World and family farmers in the First World.

Background to the international food regime

An earlier international food regime had existed between 1870 and 1929, when millions of European migrants settled the great wheat-producing areas of the world – the American plains, the Canadian prairies, the Argentine pampas and large areas of Australia – to supply grain to the growing working classes of Europe. While European production was sustained on the whole (except in England), imports from these new areas increased sixfold between 1870 and 1929.[2]

The old regime collapsed during the Great Depression and World War II. Large numbers of farmers in the new grain areas faced crisis when they lost their export markets in 1929. Subsequently European states were determined to end their dependence on grain imports, and to shore up their own agriculture. Despite the problems for American farmers, US policy supported the turn to European self-sufficiency in wheat. Beginning in 1947, the American government sponsored substitution of its own exports in Europe directly through Marshall Aid. A decade later it supported the Common Agricultural Policy of the European Community – the very policy it now attacks.

The loss of the European market was a problem the United States shared with the other major grain exporters, but its position was different for two main reasons. First, the United States had become the leading world power and the dollar was the new world currency: this gave the American government possibilities not available to other countries for financing international trade.

Second, the United States government held growing wheat surpluses as a result of the farm policies of the New Deal. At the end of World War II, only the United States had policies supporting agricultural prices through government loans which farmers paid off in grain. The Agricultural Adjustment Administration (AAA) and the Commodity Credit Corporation (CCC) had been created in 1933 as the first major act of the New Deal of the Roosevelt administration. This reflected the political strength of farmers in the coalition supporting the government. Such farm policies (with modifications) remain politically important to this day. Although other advanced capitalist countries eventually adopted policies creating government-held surpluses, at that time American surplus stocks reflected a unique combination of farmer strength in national policy and national strength in the world economy.

A third element in the new international food regime was the decolonization of Africa and Asia. Most were agrarian societies, self-sufficient in food grains. The countries of Latin America, often more industrial, were also generally self-sufficient in food, and some were exporters. Politically, as the Cold War developed, they were a challenge to American incorporation into the free world; economically, foreign aid could facilitate capitalist industrial development. Food aid played a particular role in shaping the proletarianization of peasant societies and regions – in Marxist language, in primitive accumulation.

A fourth element was the Cold War itself. Trade embargoes ruled out the socialist countries as markets. In addition, by the late 1950s the Soviet Union had temporarily recovered its historic role as a wheat exporter. It shifted to a wheat importer as *détente* allowed, because domestic grain production could not keep pace with the shift to more meat in national diets.

Food aid and the international food regime

Food aid was the combined solution to American surpluses and to further integrating Third World agrarian societies into the capitalist sphere of the world economy. It was the pivot of the international food regime of 1947–72. In the 1950s and 1960s bilateral arrangements between the US and Third World states became the typical transaction of the international wheat trade.

The cumulative effect of these deals was to change the pattern of international trade. In the old regime, the US was one of several exporters all depending on Europe as a market. In the emerging regime, the US became the dominant export country, and a multitude of Third World countries came to depend on subsidized grain imports.

Food aid was the mechanism of this shift. Although it was to play an important role both in capitalist development (or underdevelopment) and in the military projects of the Cold War, no one foresaw these possibilities in their entirety. At first food aid was the result of attempts to solve the American farm surplus problem and to reconstruct Europe and Japan after World War II.

Ironically, food aid was thus invented in its modern form (loans in non-convertible local currencies) to solve the problems of the post-war age in *Europe*, under the Marshall Plan to Europe between 1948 and 1952. This project was designed to reconstruct the European economies within a capitalist world economy open to the flow of goods and capital, with payments in US dollars. At the same time, Marshall Aid was rightly seen by many of its admirers and critics as a key step in creating the hostile blocs of the new Cold War. On one side, the Soviet Union refused to participate (reflecting realistic concerns about American intentions to restore capitalism), and on the other, NATO was formed in 1947, creating a military and political backdrop to the economic embargo of the Soviet Union.

The economic obstacle to an open capitalist world economy was the inability of war-torn Europe to pay for anything, from food to factory equipment. Europe lacked foreign exchange, which in the Bretton Woods system meant dollars (exactly the problem faced by Third World countries needing to import goods for industrial investment). For Europe, it was solved through US aid, including a large component of food aid.

However, while US food aid helped to build European self-sufficiency in wheat,[3] its effect in the Third World during the next two decades was to be the opposite. It undermined local agriculture, creating new proletarians dependent on commercial food, and new nations dependent on imports.

How food aid worked

Food and agriculture accounted for 29 per cent of Marshall aid, or about $4 billion between 1948 and 1952. In this way the US was able to subsidize a ninefold increase in exports between 1945 and 1949 to a level higher than before the war. In 1949 the US also began to subsidize exports under the International Wheat Agreement. By 1950 aid funds financed over 60 per cent of all American agricultural exports. These sales were made at negotiated prices which averaged 62 cents a bushel less than the supported domestic prices. The government carried the charges for this, to the sum of $2 billion between 1932 and 1953. Meanwhile, despite the absorption of much of the government grain by the Korean War between 1950 and 1953, and despite continuing attempts by the AAA to control output through limiting acreage, surpluses continued to mount. The end of Marshall Aid and the Korean War thus created serious problems.

For their part, Third World countries faced historically unprecedented dilemmas of capitalist development. In the global context of large-scale, mobile manufacturing capital, they could not easily repeat the strategy of many European countries and Japan – to import food and pay for it with industrial exports. Nor could they easily repeat the strategy of the US, Canada and Australia, in which national industrial development took place through the impetus of export production of food.[4]

American food aid both solved the US surplus problem and presented an irresistible opportunity to Third World governments to overcome foreign exchange barriers to industrialization. Public Law (PL) 480, the Agricultural Trade Development and Assistance Act of 1954, drew on the experience of Marshall and other *ad hoc* aid. Title I of PL 480 (and section 402 of the Mutual Security Act of 1954) provided for sale of surplus agricultural stocks for foreign currencies. Other titles of the Act allowed for grants and barter for strategic raw materials, but Title I sales rapidly came to dominate aid shipments. Between 1954 and 1977, Title I sales accounted for 70 per cent of food aid, most of it wheat.

Title I worked through "concessional sales" at negotiated prices in the currencies of recipient countries. By the late 1970s, the US had made concessional sales under Title I of agricultural commodities valued by the US Department of Agriculture at almost $21 billion.[5] This meant that the US government paid this amount to private grain companies to ship the grain. The recipient countries in turn placed an equivalent amount of their national currencies at the disposal of the US government, set at the nominal rate of exchange of the local currency against the dollar. Since the local currencies were not convertible, these "counterpart funds" could be spent only within the country, and were available for the US government to build dams or roads, to buy supplies for military bases or any other projects that involved locally produced goods and services. It had a multiplier effect on American foreign aid appropriated by Congress, and it could be directed to projects independent of Congressional approval.

By 1956, food aid constituted almost half of all US economic aid. PL 480 aid was restricted to "friendly" countries and to those not competing with American agricultural exports. In the early years, 92 per cent of the counterpart funds held as a result of concessional sales were used as loans to recipient governments for development projects approved by the State Department, for payment of US obligations to them, and for procurement of military equipment, materials and facilities. In sum, in return for giving up dollar payments for exports (which might never have existed anyway), through Title I food aid the US got local currencies with which to pursue strategic economic and military objectives.

This was appealing to Third World governments. US food aid, working through the relation of the dollar to local currencies, got round the problem of choosing between difficult alternatives. The experience of the European Community shows that investment in national food production is expensive to governments and means high food prices for consumers. On the other hand, importing food commercially would have used scarce foreign exchange without contributing to future income through investment. Instead, most Third World countries adopted cheap food policies (often accompanied by direct public subsidies) as a way to keep wages low and so facilitate industrialization. Imports subsidized by the US made this possible with a minimal drain on foreign exchange.

Underdevelopment and proletarianization

The international food regime contributed to a new international division of labour. To indicate the role of aid in restructuring trade, American aid alone accounted for about a third of world trade in wheat in the early years of PL 480, between 1956 and 1960. Clearly, US aid was leading world trade in wheat. During the first decade of the programme, total world exports increased by more than 50 per cent. Aid accounted for about 70 per cent of US wheat exports, reaching 80 per cent in 1965. With this subsidy, the US share of world trade increased from just over a third prior to PL 480 to more than half in the early 1960s.

Since so much of world wheat trade hinged on American aid, it is not surprising that the recipient countries together became the major importers.[6] The underdeveloped countries of Africa, Asia and Latin America[7] rose from being practically non-existent as importers to taking almost half of world imports in 1971 – and at their peak in 1978, they bought 78 per cent of American wheat exports.

The effects on domestic agriculture could be extreme. The best-known case is Colombia. According to a well-known study, after fifteen years of concessional sales, imports of wheat had increased ten times, from almost 40,000 tons in the early 1950s (before PL 480) to almost 400,000 tons in 1971.[8] Concessional sales accounted for 53 per cent of imports until 1962; after 1963, when the market had been "developed", sales could continue to increase without concessional arrangements. Aid – now in the form of dollar credits – fell to 30 per cent of imports. The results were dramatic. Between 1951 and 1971 domestic prices were cut in half and wheat production fell by almost two-thirds. Domestic wheat production went from 78 per cent of consumption to 11 per cent. At the same time, other food crops, such as potatoes and barley, went out of production too. The authors of the study conclude that the collapse of agriculture meant that former peasants became marginally employed or unemployed.[9]

The international food regime of 1947–72, therefore, contributed to a typical Third World pattern of rural underdevelopment and dependence on food imports. *Within agriculture* the underdevelopment of the food sector was parallel to the continuing transformation of agriculture for export. The people expelled from the land due to capitalist export production turned to urban food markets to buy what they needed to eat. The import of cheap food meant that these markets were supplied from abroad. Small food producers could not compete with dumped American wheat, which was often subsidized further by the national government.

To the spectacular growth of urban populations, it contributed food at world prices depressed by the chronic surpluses held (and periodically dumped) by the US government, and often directly subsidized as well. This makes sense of the surprisingly consistent national policies favouring cheap food as a means to rapid industrialization. It meant a shift from domestic cereals to imported wheat. By the early 1970s, imports came to account for 26 per cent of Third World wheat consumption, compared to less than 1 per cent of other cereals consumed.[10]

Thus wheat was a vehicle of proletarianization. It marked a change in diet for new participants in food markets. Annual per capita consumption of wheat increased by 69 per cent (from 31.6 to 53.5 kg) while per capita consumption of other cereals increased by only 17 per cent (from 138.7 to 162.6 kg), and per capita consumption of potatoes and other root crops actually declined by 21 per cent (from 109.2 to 86.2 kg).[11] Wheat was still only 25 per cent of all cereals consumed at the beginning of the 1970s (compared to 18.6 per cent in the early 1950s), but as the case of Colombia suggests, it was the specific food staple representing a shift to *commercial* food. Combined with surplus labour from capitalist and underdeveloped agriculture, subsidized imported wheat meant very low wages for those who found employment at all.

The end of the international food regime

This international food regime depended on an American monopoly of surplus wheat disposal through special currency arrangements. It started to collapse in the late 1960s with the decline of US surplus stocks, the rise of the world wheat price, change in the status of the dollar, and the participation of other countries in foreign food aid. An era ended decisively in 1972–3 when all surpluses temporarily disappeared with the Soviet-American grain deals marking *détente*. This end to the mutual economic isolation of the capitalist and socialist blocs had been anticipated by Canada's exports to China in the late 1960s, but the Soviet-American pact became the largest single transaction in the international grain market.

Surpluses are now held by several major capitalist countries, at a time when the US has become a net importer of all foods, and needs to export grains more than ever. Trade wars are looming between the European Community and the United States, while Canada, Australia and other grain exporters suffer as non-combatants. Direct attacks on each other's markets by the USA and EC is but a new twist in the larger competition for world markets.

This competition has extended to the socialist countries. Of course, the Third World remains a field of competition, with special credit arrangements as a major weapon. Yet the focus is now clearly commercial. The basis for this change is the *success of American aid programmes in developing Third World markets*. The hungry masses of the Third World are now potential consumers of surpluses that are increasingly international.

The result is clear in the separation of grain trade from food aid. In the late 1960s the amount of US food aid fell by a third, while US exports declined slightly. Commercial sales were replacing aid, and other exporters were able to compete both commercially and as aid donors. Thus, as a proportion of US exports, US aid fell from 71 per cent in the early 1960s to 49 per cent at the end of the decade; as a proportion of *world* exports, US aid fell from 36 to 17 per cent in the same period. The US share of world exports meanwhile fell from 50 to 34 per cent. In the early 1970s, when the Soviet–American grain deal tipped the balance, the US export share increased again (to 40 per cent) but US aid plummeted to 13 per cent of US exports and 5 per cent of world exports, which is roughly where it remains.

Aid has become increasingly multilateral and the grant component has increased. As other advanced capitalist countries became donors, by 1973 the total world food aid of $1.2 billion had a multilateral component of $277 million or about 22 per cent. The US share of world food *aid* had been over 90 per cent throughout the 1960s, but in 1969 it fell to 82 per cent and by 1973 had fallen to 59 per cent. In 1973, a *multilateral* component – donated to the World Food Programme of the United Nations or though other channels – accounted for 7 per cent of US aid. In addition the *grant* component of bilateral food aid increased. In 1973 it accounted for 37 per cent of American bilateral aid, and 45 per cent of all bilateral aid by all donors. As food aid shifted away from trade, donors came to include countries without surpluses, such as Great Britain (at the time), and even countries that imported grain, such as Japan.

Even though the political shift corresponded to new economic realities, it was by no means automatic. Paradoxically the political and strategic aspects of aid came to dominate just at the time when it was achieving economic success. During the Indochina War food aid had increasingly been directed to Vietnam and Cambodia. Much of the local currencies held by the US in exchange had been reloaned as economic and military aid. This effectively increased US military resources beyond those allocated by Congress. In the aftermath of Watergate, amendments to existing legislation prohibited "mutual defence" uses of food aid funds. Building on earlier amendments encouraging a shift to commercial sales and grants, they assigned priority to grants and to countries defined as poorest by the United Nations.

Reduction in American aid also presupposed a decline in the political strength of wheat farmers. As virtually every commentator on the domestic effects of price supports for farmers has noted, this type of subsidy rewards the largest farmers most. It has encouraged farms to grow larger and to decline in number. The number of people working in agriculture fell by more than half between 1950 and 1972, and their proportion in the total labour force fell from 15 to 5 per cent. In the 1970s boom that marked the end of the international food regime, farmers expanded as fast as they could through borrowing. When prices fell again, they faced the same problems as before, now intensified by a crushing debt load. As their numbers have declined, pressures to reduce government expenditures have increased. Despite the counter interest in expanding agricultural exports, the end of New Deal-type farm programmes may be imminent.

What now?

The international food regime created complicated international class relations. On one side, rapidly declining numbers of family farmers in advanced capitalist economies depend on exports to rapidly expanding Third World masses who must buy whatever food they eat. On the other, consumers in advanced economies depend on food exports from the Third World. Workers depend on cheap meat products from land once used by peasants, and privileged consumers benefit from the "strawberry imperialism" that devotes large fertile areas and outrageously low-paid labour to supply exports of fresh and preserved fruits and vegetables.

The contradictions are more complicated than in earlier food politics because they are so highly mediated by transnational relations. Even when food supplies were mainly national, politics revolved around alliances that cut across class conflicts. For instance, through the Anti-Corn Law League in early nineteenth-century England, capitalists and workers joined to oppose tariffs protecting the landed class. While cheaper food would benefit "the nation" as opposed to a sectional interest, it would also mean lower wages. Capitalist interest hid beneath the struggle against remaining precapitalist privilege. Late nineteenth-century Germany presents a contrasting example of how sectoral politics can overwhelm class politics. Through the Agrarian League, the large landowners of eastern Germany mobilized the small farmers of western Germany to support the tariff. This set back the commercial development of food production by the small farmers along the lines, for instance, of Denmark, creating a dependence of small farmers on the landed class. The ultimate effect was to foster extreme right-wing political sympathies among them.[12]

As the connections of the post-war international food regime are unhinged, the *national* strategies pursued by governments hide the *class* dimensions of the problem. All major export countries seek to solve their national farm problems by competitive dumping on Third World (and socialist) markets. It is a counter-productive strategy for the remaining First World farmers. Increasing poverty and increasing agricultural production in the Third World have already set limits to market expansion there (limits reinforced by austerity measures imposed by international agencies). The socialist countries are benefiting from subsidized imports of grain and dairy products, but neither farmers nor workers in the capitalist world profit from this trade.

The majority of Third World states are promoting capitalist strategies of domestic food production. Reinforcement comes from the banks and the IMF seeking to reduce Third World imports and state expenditures where they can. Radical solutions that expect peasants to remain self-sufficient hold no hope.[13] The hungry masses must buy food at low prices. If it is not to come from abroad it must come from national farmers who must increase productivity.

This is already happening in a capitalist form. The history of agriculture in international capitalism is one of impoverishment of direct producers, though to different degrees and in different ways, and the degradation of the natural resources which are our common heritage. The tendency now is to reproduce the capitalist experience of over-specialized agriculture and standardized processed edible commodities on the vast canvas of the Third World.

For advanced capitalist countries, the decades after World War II marked a drastic change in farming and in diets. Fewer farmers used more machinery and chemical inputs on larger and more specialized farms, leading to depopulation of the countryside, destruction of rural communities, and ecological damage. They none the less became smaller links in increasingly complex agro-food chains. Mass-produced food became an important industrial and retail sector, mediating between highly specialized farms and consumers of thousands of new food products (about 12,000 items are sold in a typical supermarket).

The Green Revolution is reproducing in the Third World the experience of the US "corn revolution" – increasing capitalist control over mechanical, chemical and genetic inputs, and the manipulation of plants and animals to maximize durability and transportability over long distances. As people in the underdeveloped parts of the world become dependent on commodified food, their diets are ruptured from local ecology and tradition and restructured through international food markets.

Alternatives

The alternative is to reconstruct agriculture within a socialist project of local, regional and national planning, and ultimately international co-ordination. This applies to rich and poor countries alike. The specialization of farms and of countries in particular products is so extreme that it seems impossible to understand, let alone control, the production and distribution of various foods. Local and national control requires that consumers and producers everywhere work to undo this extreme specialization, which makes us all vulnerable to agro-food capitals, to our mutual (though unequal) loss.

For the Third World, Soviet-style collectivization is not a path to democratic socialism. A general lesson is that agriculture must be invested in rather than accumulated from, whatever the mixes of scale for various plants and animals. A specific lesson for areas with small producers is that their spontaneous participation is vital to the economic success of a socialist restructuring of agriculture. Family labour can be a vital part of collective agriculture for as long as people wish to have it. Hungary is a case which deserves study because, on the positive side, it realizes the economic potential of "private" participation in planned agriculture; negatively, it sets limits to egalitarian participation in the overall direction of co-operatives and state farms.[14]

The capitalist division of labour and the forced collectivization in Soviet state socialism offer two negative lessons. Repelled by both poles, direct producers and consumers must find a path to a democratic socialist agriculture appropriate to each locality, region and nation.

Socialist food politics, then, are not about self-sufficiency as such, but about *self-determination*.[15] Imports and exports – as Cuba and Nicaragua have learned – can play a role, depending on the relation of national resources and needs. There is no standard solution to the balance between local production at any level and trade. What is crucial is the collective, popular political will to take control of local and national resources and democratically determine needs. Only then will the international economy serve rather than dominate production and consumption.

These lessons apply to the advanced capitalist countries. We do not know the social cost of the food we eat because it is not included in the price. This is true even for domestic foods, such as wheat, whose true cost will be known only when the effects of unsustainable agricultural practices on natural resources and human health begin to be counted as part of the food system. Family farms have been forced to continue competitive specialization. In consequence, most of them have disappeared, and the rest are locked into integrated agro-food chains. No more than the self-supplying peasant can specialized, capital-intensive, competitive family farmers be "saved".

The costs of the food we import from the Third World are hidden in more complex ways than those of our exports to them. Large areas in Latin America are devoted to supplying fresh and processed fruits and vegetables to relatively privileged consumers in North America and Europe, and similar transformations are occurring in Africa to supply the European market.[16] Third World export production is undertaken by agro-capitals precisely because land and labour are valued so low.

The low prices at which these exotic imports are offered to consumers undercuts efforts in advanced capitalist societies to reorganize food production and consumption as an aspect of the complex needs and resources of the community. For example, while American workers eating hamburgers at a fast food restaurant are better off than the Latin American peasants expelled from the land on which the beef grazes, both are damaged by the agro-food capitals connecting them. Now McDonald's is expanding its sales in the Third World, to those who can afford to buy.[17]

For the poor in rich countries, cheap hamburgers – and fast food generally – are a rational diet in a life whose work and domestic relations are structured by late capitalism. Changing this aspect of exploitation means making the connections between attempts to control our own lives, and struggles by local and national communities in the Third World to regain control over the land now appropriated by capital, for example for export production of cattle.

It is in the interest of socialists everywhere, including those privileged with ready access to the produce of the globe, to have all agricultural products reflect their real human and environmental cost. (Of course, socialist communities may wish to subsidize these costs to individuals.) The only way for real costs not to be borne by impoverished agricultural workers and desperate small farmers is for standards of living to be raised. And agricultural workers must participate with consumers within democratic, national frameworks collectively to determine the uses of the land.

What would socialist agriculture be? To begin, it would have to reconstitute both depopulated areas of monocropping and denatured suburban and ex-urban settlements. The renewal of social life and natural processes involved in the production, distribution and consumption of food means more localism. Those who produce food and those who eat it need to be in more direct contact with each other, to redefine needs (for farmers, food workers and consumers) to include a renewed connection to our bodies, to each other and to the species we cultivate and consume.

The deepening of our humanity lies in redirecting our capacities as part of developing our needs. This must include an enlivened experience of ourselves as part of the natural world. Before capitalist restructuring drives out all remaining communities still intimately bound to their local settings, we must try to recover their knowledge and understand their practices. This is the agenda for a new science serving a renewed society of free individuals. It is being anticipated by farmers and agronomists at the margins of agriculture.[18]

Our apparently individual choices are really based on the relative wages and prices created by agro-food capitals, as they continually restructure regional production and global consumption. The question for socialists is whether local communities (in mutual co-ordination) will reorganize the world of food locally, as part of new social relations respecting and embracing the natural world we inhabit. As local struggles over land and labour begin to make the prices of commodities everywhere reflect the real human and natural costs of their production and transportation, we shall be better able to develop our understanding of the real relationships connecting us to the producers of what we eat and the customers of what we make. Then we shall be able to take from capital the power to shape our social relationships and our life experiences.

Notes

1. This chapter summarizes sections of my forthcoming book, *The Political Economy of Food* (London: Verso/New Left Books).
2. When not otherwise specified, all sources can be found in my earlier article, "The political economy of food; the rise and fall of the Postwar International Food Order", in the Special Supplement to *American Journal of Sociology*, vol.88 (1982), entitled *Marxist Inquiries* and edited by Michael Buraway and Theda Skocpol.
3. To a large extent the US made up for its lost European wheat sales through exporting feed to the animals in the growing meat and dairy sectors of Europe. Hybrid corn and soybeans became major components of US agriculture, supplementing the historic mixed corn/livestock and wheat production of the plains.
4. For this specific path of capitalist development, based on industrialization within an agro-food complex, see Charles Post, "The American road to capitalism", *New Left Review*, no. 133 (1982). Earlier arguments along the same lines are much more frequent for Canada. For a classic example, see Vernon Fowke, *The National Policy and the Wheat Economy* (Toronto: University of Toronto Press, 1957).
5. See the table reproduced by Mitchel B. Wallerstein, *Food for War – Food for Peace* (Cambridge, Mass.: MIT Press, 1980), p. 53. Wallerstein's book is a good source of detailed information on American food aid and the debates surrounding it.
6. Japan, which accounted for 5 per cent of world wheat imports in 1959 and 9 per cent in 1972, was the other major import country to emerge within the international food regime.
7. Excluding Argentina.
8. Dudley, Leonard, and Roger Sandilands, "The side effects of foreign aid: the case of Public Law 480 wheat in Colombia", *Economic Development and Cultural Change*, vol. 23, no. 2 (1975).
9. Economists generally agree about these effects on *agriculture*. There is disagreement about how to evaluate the net *welfare* effects of foreign aid. In other words, aid generally and food aid particularly may have had positive effects compensating for the

negative effects on agriculture. From a Marxist perspective, this sort of measure is less important than the restructuring of the economy, that is the contribution to the expulsion of people from the countryside and the formation of urban and rural proletariats.

10. Consumption is calculated as production plus net imports or minus net exports. Third World includes all countries of Asia (except Japan and China), Africa (except South Africa) and Latin America (except Argentina) and the Caribbean. Calculated from FAO *Trade and Production Yearbooks*.
11. All data calculated from FAO *Trade and Production Yearbooks* and UN *Demographic Yearbooks*.
12. One good discussion of the Anti-Corn Law League is Reinhold Bendix, *Work and Authority in Industry* (California, 1974). The classic study of the Agrarian League is by Alexander Gerschenkron, *Bread and Democracy in Germany* (Berkeley and Los Angeles: University of California Press, 1943), and the special response of Denmark to the late nineteenth-century fall in agricultural prices is argued by Charles Kindleberger, "Group behaviour and international trade", *Journal of Political Economy*, vol. 59 (1957).
13. See Susan George, *Stratège de la faim* (Geneva: Editions Grounauer, 1981).
14. Swain, Nigel, *Collective Farms Which Work?* (Cambridge, 1985).
15. Moore Lappé, Frances, and Joseph Collins, *Food First* (New York: Houghton Mifflin, 1977). See also George, op. cit.
16. Burbach, Roger and Patricia Flynn, *Agribusiness in the Americas* (New York: Monthly Review Press, 1980), and George, op. cit.
17. Feder, Ernest, "The odious competition between man and animal over agricultural resources in the underdeveloped countries", *Monthly Review*, vol. 3, no. 3 (1980). On Third World meat consumption, see Prof A. Yatopoules, "Middle-income classes and food crises: the 'new' food-feed competition", *Economic Development and Cultural Change*, vol. 33, no. 2 (1985).
18. For example, Wes Jackson and Marty Bender, "An alternative to till agriculture as the dominant means of food production", in Lawrence Bleech and William B. Lacy, *Food Security in the United States* (Boulder, Col.: Westview Press, 1984).

3 Moving the Lever: a New Food Aid Imperialism?

Ben Crow

Food aid is rarely given to one government by another without conditions. Donor governments usually expect to influence the political or economic policies of the recipient through their gift. Where food aid supplies a significant proportion of the food consumed within a country, the influence of the donor may be correspondingly large in the determination of government policy.

Through an examination of the relationship between the largest food aid donor, the US government, and one of the largest and poorest of food aid recipients, Bangladesh, this chapter describes some of the institutions and processes of leverage which have been established by food aid donors.

During the 1950s and 1960s, US food aid was driven by domestic surpluses, and the conditions which recipient states had to fulfil derived primarily from US foreign policy (see Friedmann's chapter in this volume). This phase ended in the late 1970s with a series of Congressional amendments which intended food aid to encourage economic development and equity. US economic and foreign policy objectives remained overriding factors but the detailed determination of conditions was to be guided by the promotion of development in recipient countries. This second phase may now be evolving into a third. In 1989, a new round of Congressional discussions led by Congress Asia Committee Chair Stephen Solarz sought to link aid to the establishment of democratic institutions in developing countries.

If we define imperialism as the process by which the

dominant social group in one country influences the social order in another, then there is no doubt that food aid conditionality is imperialism. The second phase of US food aid conditionality, however, appears to be a benign use of power. Separating ends and means reveals some of the complexity of this new imperialism. The ends include economic growth and some elements of equity as well as the promotion of (a particular variant of) capitalism. The means are the exertion of economic and political power without democratic check or balance.

Food aid conditionality in fact is part of a wider process of global economic and political integration led by the dominant capitalist powers, in which IMF and World Bank "structural adjustment" is another element (see Mbilinyi's chapter in this volume).

We can identify different forms of this integration. In one, officials from the dominant capitalist power move directly into positions in Third World states. This has occurred in Liberia, where fifteen US economists have been inducted into the government with expenditure-authorizing powers and deputy minister status. Though members of the Liberian government, they are responsible to a US team leader and remain on the US government payroll. This extreme, with its parallels to nineteenth-century imperialism, reflects the ambitions of some of the international and US officials who provide advice to Third World governments. Often frustrated by the weakness of the states they advise, and the corruption and lack of co-ordination that hamper the implementation of their advice, they would love to have the ability to implement their proposals directly.

This complete integration seems to be rare. In another more common form, the dominant capitalist power or powers are not inserted inside government but control policy making and part of the state's expenditure in one area of the economy. This may be a more or less formal arrangement. In Mali, for example, it is relatively formal. All money generated by the sale of food aid is placed in a common fund. Expenditure from the common fund is authorized by a committee of donors with representation from the government of Mali. In this case, the US government notes that it

> benefitted from the existence of a multi-donor effort to influence the government of Mali to liberalize cereal marketing policy. . . .Membership in this group enabled the US Ambassador to have increased influence in the policy dialogue arena over and above what might otherwise have been possible. . . .[1]

Food aid conditions have spawned a complex growth of institutions, processes and relationships which allow the US government to influence the management of recipient economies. Amongst many "successes", USAID claims its use of food aid leverage has: determined agricultural policy in Pakistan and food policy in Bangladesh; guaranteed the Camp David accords; and sustained and influenced government in Sri Lanka, Peru, Jamaica and Haiti.[2] The 1977 Congressional amendments which provided greater force to the development objectives of food aid also consolidated the institutions of leverage, and focused them on changing the policies of recipient governments. Moreover, other food aid donors, such as Canada, the European Community, Britain and Australia, are following the US example by using their donations to change policy along the same lines. USAID studies emphasize the importance of donor co-ordination for the success of policy leverage.

This chapter examines the influence of food aid conditionality in the case of Bangladesh, where the contrast between the two phases of US food aid conditions is particularly stark. In 1974, the denial of US food aid was a factor in economic and political instability which led to famine and the downfall of the country's first government. Since 1977, the new institutions of food aid conditionality have been used to determine food policy, and to improve the Bangladesh government's ability to avoid famine. By doing so, they have considerably strengthened the current government. The contrast then is between outcomes; the methods of leverage have been used consistently and have grown stronger.

Bangladesh is one of the largest of the very poor countries. It is the second largest recipient of food aid, and nearly half of all government expenditure comes from foreign aid of all kinds. Aid has taken the dominant place once held by land taxation in government revenue. Its level and continuity are a central concern of government. Food aid alone contributes approximately 13 per cent of government resources.

The famine of 1974

In 1972–3, the US supplied 500,000 tons of food aid to Bangladesh. In March 1973, negotiations started for the following year's supply, but stalled: the US State Department wanted changes in food policy.[3]

No new US food aid was committed until early 1974. Then, before a commitment to supply 150,000 tons could be completed, the US government decided that Bangladesh's export of jute sacks to Cuba contravened a section of US food aid law which forbade aid to countries trading with communist governments. This happened just as rice prices were reaching a peak and government stocks had reached a very low level. At this point the sluggish negotiations may have became a formal embargo.

No US food aid arrived until late 1974. Famine deaths had reached a peak, at which they were to remain for most of 1975. A combination of economic instability (leading to rapid inflation), fears about government food stocks, reduced harvests in the wake of floods and widespread political collapse caused a threefold rise in the price of rice during the first half of 1974. Agricultural labourers found their food purchases squeezed between rising prices and falling wages. A shaky government, with falling stocks and insufficient foreign exchange to buy food, cut back on food distribution to the countryside in order to supply the cities.

During the twelve months leading to the peak of the famine, US food aid amounted to 17 per cent of the tonnage supplied in the previous year, and overall food imports were down by nearly 1 million tons. The cut in US food aid was approximately 400,000 tons, similar to the cut in food distribution to rural areas.

The US agenda for reform of food policy

In 1975, US advisers in Dhaka formulated a food policy agenda. In contrast to the lack of interest shown by Washington during 1973 and 1974, this agenda reflected a growing concern about food insecurity. The most detailed account of the US food policy agenda has been published in a book by one of those who contributed to its formulation.[4] The essentials of the agenda were:

- greater reliance on private foodgrain markets
- reduction of the public food distribution system
- termination of subsidized food to urban populations
- introduction of an open market sales price-stabilization scheme
- generally higher food prices.

The process of leverage

There was a second break in food aid supplies in 1976, intended to get the agenda of reforms moving, and there are reports of further breaks in subsequent years. In general, however, the introduction of the US agenda has not required overt expressions of political power on the scale of 1974. Mostly, "reliance on relations of influence", associated with the negotiation of food aid agreements, has been sufficient.

The key US legislative instrument, which sets the context for these negotiations, is known as Title III of Public Law

480. Title III food aid is desirable from the point of view of the recipient government because it provides free food on a multi-year basis, but it also has more binding conditions.

The recipient government is caught in a complex double bind. For the food aid to be supplied free they first have to undertake policy reforms negotiated with the US government. Then they have to use the proceeds of the sale of US food in ways agreed with the US government. If they fail to comply with either condition, they face one of two penalties: first, "loan forgiveness" may be denied, that is the country may have to pay for the food in years to come; second, behind all such negotiations lies the threat that US food aid could be delayed or cut off.

Negotiations: "more policy bang per buck"

In order to get a commitment for a steady supply of US food aid, the recipient government has to enter into negotiations, initially for a multi-year agreement, which gradually become an institutional feature of relations between the two governments.

In principle, agreement emerges from "policy dialogue" between the half-dozen co-ordinating and line ministries on the Bangladesh side and the four or five departments representing the US. In practice, US representatives make political choices, backing the policies, agencies and individuals they like, and levering out those they don't. A "review of successes" published by USAID frankly advocates the provision of funding to recipient government agencies forced by the "dialogue" to make sacrifices: "Where such 'sacrifices' are backed up by jointly programmed local currency sales proceeds, both negotiation and subsequent implementation tend to be more successful."[5]

The first ten years of policy dialogue in Bangladesh have been dominated by the 1975 US agenda. With each new multi-year agreement, US consultants suggest new policy reforms. Recent agreements encourage the introduction of maize production (as earlier agreements had introduced wheat consumption and production) and the privatization of cotton and oil imports, but the general focus and coherence of the policy agenda remains that framed in 1976.

Negotiations cover detailed issues of policy implementation as well as the broad direction of policy. The first agreement with the Bangladesh government included the text of a four-page circular, which the Food Ministry agreed to send out to its local offices, detailing the exact procedures for carrying out open market sales (OMS) of foodgrain from the public food system. Successive agreements[6] have become longer and more detailed, and those drafting them on the US side have become more skilled. A Washington USAID official described the 1987 agreement as giving "more policy bang per buck" than its predecessors.

The "policy dialogue" does not end when the food aid agreement is signed. The agreement contains conditions for the recipient government to meet, reporting requirements, and the expectation of annual evaluations and periodic renegotiations. There were at least thirteen separately negotiated amendments to the 1982 agreement, and the 1987 agreement had four by early 1989.

The US government does not always fully achieve its ends in these negotiations. In those during the implementation of the second agreement (1982–6) and the run-up to the third (1987), the most contentious US demands were for the abolition of the public food distribution system, of the subsidy on fertilizers, and of subsidized food for the security forces. There was some compromise on all three, but the US government achieved the spirit, if not the letter, of its reform programme in every case. As one former Bangladesh Agriculture Minister estimated, the food aid donors, led by the US, determine "70% of food policy, and their counterpart funds give them a handle on agricultural policy too". Some elements of Bangladesh food policy are now monitored more closely in the US Embassy in Dhaka than in the Bangladesh government, enabling US advisers to "kick ass" if their policies are not followed.

For food aid to keep coming, the policy reforms have to be implemented. For "loan forgiveness", the funds or the food have to be used in ways agreed with the US government.

Counterpart funds and donor co-ordination

Most Title III food aid is sold by the Bangladesh government through its ration distributions or through open market sales. It therefore generates large local currency funds. The food aid agreements determine how the funds should be spent. By 1987, $250 million had accumulated. (A US evaluation mission was concerned to find that it had been lost in a ledger of the Bangladesh government.) As Gramm-Rudman legislation (imposing expenditure limits on the US government, in an attempt to reduce the US fiscal deficit) reduces the amount of money available for US foreign aid, these local currency funds have increasing importance. AID notes, "greater attention is being given to local currency programming since from 1985–7, it is anticipated that annual expenditures of local currency [worldwide] will be about $2 billion."[7] US AID in Dhaka has $80 million a year (in local currency) to spend from food aid sales.

In recent years, US representatives have spent 60 per cent of this money on water development projects, and 25 per cent on government budget support. The 1987 agreement directs nearly $3 million for a team of economists from the International Food Policy Research Institute in Washington to back up the work of a government agency, the Food Planning and Monitoring Unit (FPMU), itself created by the 1978 agreement. The FPMU gave the US government access to food statistics, and US advisers another direct line to the food policy process.

In the case of Bangladesh, the institutions of donor influence have not achieved the formal integration of the donor committee in Mali or the US fifth column in Liberia. They are, nevertheless, numerous and growing. Several donor committees meet regularly to co-ordinate aid and discuss policy, and there is agreement about the division of policy leadership between donors: USAID takes food policy, the IMF (and latterly the World Bank) look after macroeconomic policy, the World Bank provides the lead in the assessment of development projects, and so on. The conditions set by the lead donors for policy in each area are incorporated into the aid agreements of other donors.

In recent years, this co-ordination has been subject to some strain. A group of donors, led by Scandinavian governments and calling itself the "like-minded group", has attempted – unsuccessfully – to establish an alternative development agenda, emphasizing the alleviation of poverty rather than the spread of capitalism. The European Community, disposing of its large surpluses, has now also begun to seek influence on food policy. Other donors, however, lack the history, expertise and personpower which USAID brings to the task. US representatives report that other donors have a tendency to line up behind their policy conditions – "that looks good, we'll buy into it."

Outcomes

The outcomes of this process occur at different levels. At the level of national economic policy, aid conditionality has helped to negotiate a shift away from state control of commerce and industry and towards greater reliance on private ownership and distribution. This was not encouraged solely by US food aid, or indeed by any one donor, but US food aid conditions took the lead by restructuring fundamental elements of food and agricultural policy. A new framework for food policy management has been introduced, which has been one factor in the weathering of crises. The new food policy has also opened up Bangladesh to the international economy, and introduced it to the consumption of wheat.

At the national political level, food aid has assured political stability. Historically, the food aid withdrawal of 1974, whatever its intentions, contributed to the downfall of a weak social democratic government, and its replacement by a more conservative military regime. When that regime came in, and when it was succeeded in the 1982 coup d'état which brought the current president to power, there was little hesitation about the supply of food or any other aid. Donors were ready not only with aid but with sets of ideas which enabled some semblance of capitalist development to occur.

The role of food aid in maintaining political stability can also be seen at the local level. Reforms of the public food system are apparently now being used to build links from the president's party to influential people in the rural areas.

As a result of the 1987 food aid agreement, the Bangladesh government is reforming its rural rationing system. USAID criticized the old system because it delivered cheap food to the wrong people, and because 65 per cent of the food leaked on to the market. The key questions about the implementation of the new system are: who gets the ration cards? and who is appointed dealer (therefore benefiting from the leaks)? Bangladesh newspapers suggest that the reform of rural rationing is being used by the president's party to create the rural base it needs if it is to hold credible elections in the future. With the substantial patronage of foodgrain dealerships, it can buy the support of local influential people and the votes they command in the rural areas.

Why has the Bangladesh government been so easily influenced by US food aid conditionality? Since 1975, aid has supported a series of conservative regimes, and conditionality has been used to encourage the influence of conservative elements within each regime. As the current president of Bangladesh made plain in his 1989 visit to Britain, he has become an admirer of Mrs Thatcher, and accepts the ideas of the new conservatism, reducing the role of the state and giving greater scope to private trade.

US food aid conditions contravene fundamental ideas about democracy and (opposition to) colonial rule enshrined in the American Declaration of Independence. Concern about this new form of imperialism is not just a reflection of the Bangladesh bureaucrats' complaint about being pushed around by American economists fresh out of Harvard. Perhaps food aid conditionality will conform to axioms of human rights only when Bangladesh peasants are given a vote in US elections.

Notes

1. USAID, *Negotiating and Programming Food Aid: a Review of Successes* (Washington: Bureau of Food for Peace and Voluntary Assistance, USAID, 1986), p. 16.
2. USAID, *A Comparative Analysis of Five PL 480 Title I Impact Evaluation Studies*, AID Program Evaluation Discussion Paper No.19 (Washington: USAID, 1983); and USAID, 1986, op. cit.
3. State Department cables released under the Freedom of Information Act. See Ben Crow, "US policies in Bangladesh: the making and the breaking of famine?", *Journal of Social Studies* (Dhaka), 35 (January 1987). A description of the context of foodgrain distribution in Bangladesh is given in Ben Crow, "Plain tales from the rice trade: indications of vertical integration in Bangladesh foodgrain markets", *Journal of Peasant Studies* (January 1989).
4. Stepanek, J., *Bangladesh – Equitable Growth*? (New York: Pergamon, 1979), pp. 76–9.
5. USAID, 1986, op. cit., p. 5.
6. US Government/Bangladesh Government 1978, 1982, and 1987 Agreement between the Government of the United States and the Government of the People's Republic of Bangladesh for a Public Law 480 Food for Development (Title III) Program. See also US Government 1988 PL480 Title III Bangladesh Food for Development Program Annual Evaluation for FY 1988, Washington, DC, and earlier evaluations.
7. USAID, 1986, op. cit., p. 8.

4 Abstract Markets and Real Needs

Maureen Mackintosh

A cautionary tale

> "Problems in food security. . . .result. . . .from a lack of purchasing power."[1]

Here is a short cautionary tale about the ways in which ideas and facts about markets are transformed and misused in the service of a currently dominant political ideology. It is the story behind the above quotation which reappears in many World Bank publications.

In 1981 the economist A. K. Sen published an important and controversial book.[2] In *Poverty and Famines* he argued that the *normal* working of markets is an important factor in the creation of hunger and famine. As markets spread through and transform rural areas, so individuals come increasingly to depend upon the workings of markets for survival, by selling goods or their own labour to buy food. The net result is an increase in the *vulnerability* of many people, especially those who own few resources bar their labour. Small farmers, pastoralists, labourers, crafts workers become vulnerable not only to drought and pests but also to changes in prices and quantities on volatile markets. Previous payments in kind are transformed into cash: "more modern perhaps, more vulnerable certainly". Old methods of insurance against disaster weaken or disappear.

As a result, famines can be caused, or more often sharply reinforced, by the normal workings of the market. The loss of crops and animals in a drought is compounded when prices of animals fall for lack of demand, and remaining food leaves a region for better prices elsewhere. Or rising prices in a boom can tip vulnerable people in poorer areas into famine. As Sen says bleakly of the 1974 famine in the Wollo region of Ethiopia, "The pastoralist, hit by drought, was decimated by the market mechanism."

Sen is making the point that these disasters are not the result of markets working badly. Markets respond to demand backed by cash, not to needs alone. Sen therefore directs us to consider the non-market determinants of the ability to command goods on the market: ownership of resources and the terms on which people come to market and which influence their ability to trade. Not people's lack of income, but in Sen's phrase, "how come they didn't have that income?" This in turn implies attention to the non-market relations which surround and structure all markets. Sen makes the point that it is social security, not high average real incomes, which prevents famine in the West: "The phase of economic development *after* the emergence of a large class of wage labourers but *before* the development of social security arrangements is potentially a deeply vulnerable one." His discussion of the famines he studied includes reflection on the need to find ways of increasing forms of "insurance" and social institutions to provide mutual support and hence to prevent, for example, competitive over-grazing of marginal land.

What happened to this important argument when the World Bank discovered it? In 1986 the Bank published *Poverty and Hunger* which referred explicitly and at some length to Sen's work, as did that year's *World Development Report*. But the argument has undergone an interesting transformation. Where Sen had demonstrated that it was not average food availability but specific changes in production and market conditions which created areas of famine even amid plenty, the World Bank states: "Problems in food security do not necessarily result from inadequate food supplies. . .but from a lack of purchasing power on the part of nations and households."[3] This tautology, poverty causes hunger, was *not* Sen's central point.

The Bank's "policy study" then proposes that "Economic growth will ultimately provide most households with enough income to acquire enough food." They then examine ways in which the hungry can be provided with a little more food in the meantime, "without waiting for the general effect of long run growth". Sen's argument that growth of average incomes alone cannot bring security has vanished. The underlying criterion for choosing methods of alleviating hunger is that they should be "cost-effective", which appears to mean "without significantly interfering with other goals such as the efficient allocation of resources". And this in turn is a code phrase for not interfering with the development of the market.

And this is the rub, for the discussion in *Poverty and Hunger* of how to interfere, minimally, in the market in order to prevent starvation was written on the rigid assumption that the liberalization and extension of the scope of markets is always and everywhere the route to both efficient use of resources and growth. The increasing integration of World Bank and International Monetary Fund (IMF) lending, and the increasingly strident insistence of the United States on the promotion of capitalism as the aim of aid-giving, has led the World Bank into a pattern of argument which begins from the assumption, largely unexplored in the documents themselves, of the superior efficiency of markets. The statements recur monotonously in the work of Bank staff and associates:

> A greater role for the private sector . . . [will] improve efficiency of domestic and international marketing.[4]
>
> These reforms [in Africa] cover a wide range of measures aimed at giving prices, markets and the private sector a greater role in promoting development in Africa.[5]
>
> The Bank . . . has consistently sought to help countries obtain the advantages of private initiative and market

> discipline as well as the benefits of soundly conceived . . . programmes beyond the scope of private enterprise.[6]

As a result of these assumptions, Sen's much more challenging agenda has to be excluded in favour of platitudes about purchasing power. And because these *are* assumptions, not conclusions, there is curiously little exploration in the World Bank's work of the functioning of actual markets: in contrast to Sen, who examines meticulously the dynamics of specific markets at specific times. *Poverty and Hunger* is rather ragged in its argument because it is hard to sustain simultaneously the view that markets promote growth by (presumably) forcing risk taking and enterprise due to the need to make profits to survive, and the view that they can promote security from hunger.

Abstract markets and real markets

When the World Bank talks about "markets" and "the private sector", there are at least three different meanings floating in the text. One meaning is the broadest abstraction, "the market", that (usually singular) token of ideological debate beloved of British and US politicians. This, usually undefined, presumably means any process of exchange, undertaken by independent actors (people or institutions), of goods, services, labour power and money. Behind it lies the concept of private ownership, or at least behaviour *as if* the items exchanged were privately owned: the Bank's (singular) "private sector". Giving this unitary label to all such exchanges implies that they all constitute the same type of process, and that an extension of "the market" is a merely quantitative change. No distinctions are made between markets for goods, labour and capital.

The second meaning of "markets" is somewhat less vague, if still abstract: it refers to the wide range of abstract "models" of different types of markets, constructed by economists for the purposes of debate; many of them generate quite different results from those attributed to "the market" in current policy pronouncements. Sen's work is an example of such abstract model building. So is the work of economists who have tried to examine the problematic theoretical links between competition, monopoly, the organization of enterprises, risk taking, investment and growth.

Finally, the phrase can also refer to the very wide range of different ways of buying and selling – for example the different types of food market – as they operate in the world. Markets in this sense of the term have widely varying institutions and economic contexts, they operate on limited information, they involve and help to create a variety of social classes, power relations, and complex patterns of needs and responses. All of this generates real effects in terms of people's survival: in short, real markets. Much of the work of analysing *these* markets has been done by anthropologists and geographers – and by market traders and marketing consultants. It is a strength of Sen's work that he also examined the details of market operation, and carefully distinguished observation from abstraction.

The World Bank's "policy studies" are difficult to read because they slide around between these three meanings. They can appear to be discussing useful abstractions or concrete processes, while actually merely invoking markets in the first general and ideological sense. Where detail slips in, it becomes clear that particular assumptions are being made about abstract models and real institutions. A good example is the Berg Report on African development which was one of the founding documents of the new wave of Bank policy pronouncements and interventions.[7] In this, it occasionally becomes clear first, that markets are supposed already to exist – they merely have to be "freed" in order to develop; second, it is assumed that what exists or will develop is highly competitive: "there should be a gradual freeing of domestic food markets to encourage greater competition. . . . The private sector, with its small-scale, decentralized and flexible structure, is particularly well suited to this task [food distribution]."

The report offers little evidence beyond casual observation for either of these assumptions. Like other Bank documents the Berg Report avoided detailed argument about which types of markets in which circumstances promote cost reduction, entrepreneurship or growth. This is curious, since the connection between atomized competitive markets of the type described in the quotation in the last paragraph (invoking as it does the economist's image of "perfect competition") and entrepreneurship of the type which promotes investment and growth is tenuous within the orthodox economic models on which the World Bank bases its work – as no less a right-wing economist than Hayek made clear many years ago. Furthermore, the report was written at a time when very little was known about the organization and dynamics of real African food markets.[8]

Real markets and real needs

The left has had trouble responding to the ideological campaign embodied in the World Bank documents. This is in part because the left has tended, like the right, to work with a dichotomy between "markets" and "planning". They were therefore uninterested in the differences among real markets, since all were seen as capitalist entities which channelled exploitation, and therefore to be abolished or at least strictly controlled. In analysing food and hunger, many on the left concentrated on establishing the bad effects of the spread of capitalist markets: an approach which produced some highly effective argument, not least *Food First* and much associated and subsequent work.[9]

These studies have traced the devastating effect capitalist development of large-scale agriculture and trade can have on small-scale food farming. They call for land reform, planning, democratic control of food production and resistance to the spread of agro-industry. The combination of evidence and accessible argument in this work has been central to the growth of understanding and activism around the politics of food in the West. It did not, however, contain much analysis of different types of markets and their effects: the studies fluctuate between a general preference for planning over markets (in, for example, *Food First*'s approving discussion of Chinese planning), and a rather unsupported view that local small-scale food markets are beneficial, while larger-scale markets are not.

History has now made this lack of clarity about markets untenable on the left. Many socialist governments in the Third World – for example in Mozambique and Nicaragua – found that where most of agriculture was undertaken by peasant farmers, markets would continue to exist. Nor were they easily controlled or regulated: although market regulation was intended to promote equality and prevent exploitation of farmers through trade, poor regulation, as the Mozambican government found, can create parallel markets and hence reinforce private accumulation and inequality. Such governments have often underestimated the class power of traders.

Furthermore, socialist governments which had reduced the scope of the market in favour of central planning of production are now re-examining the role markets should play in their economies, notably in agriculture and food distribution. The difficulties and failures of centralized planning of agriculture, and problems of food availability in the towns, have contributed to this rethinking. This in turn has made Third World socialist governments receptive to the proposition that the development of markets, required as a condition of hard currency loans, is also of benefit to their citizens, or at least the best of bad options.

The left has in consequence been forced into an overdue rethink, and is now developing a body of work on the political economy of markets which has implications for the way we think about food and agriculture. It involves bringing questions of class and power into the abstract models which we construct to help us think about markets; and it requires undertaking detailed analysis of the organization and functioning of actual markets.

The rethinking of abstract models is important because they can have a tight hold on our imagination, as the World Bank has demonstrated anew. Diane Elson has pointed out that advocates of "market socialism" tend to share with the right an image of markets as a costless and flexible method of distribution, a "form of free association".[10] Better models recognize that markets concentrate information, and hence power, in the hands of few: that some participants are "market makers" while others enter in a position of weakness; that markets absorb huge quantities of resources in their functioning; that profits of a few, and growth for some, thrive in conditions of uncertainty, inequality and vulnerability of those who sell their labour power *and* of most consumers; and that atomized decision making within a market can produce long-term destructive consequences – for example on the environment – which may have been intended by none of the participants. Finally, there is no such thing as a free market: *all* markets are structured by state action; the only variation is how the terms of their operation are set.

Studies of real food markets draw on these more useful abstractions and treat markets as sets of social relations structured by classes and institutions. Whereas left and right had tended to treat food markets as an unexamined "black box", and were interested only in production and consumption outcomes, trading is now becoming more of a subject for study on its own merits as a crucial element in economic and social transformation. Some of this work is presented in this book: between them the different chapters establish a number of premises and identify some strategic issues.

First, analysis needs to centre on the terms on which people come to market: the ways in which ownership of resources, such as land, allows individuals to establish dominant positions in organizing other markets, such as trade in food, while others for lack of resources are coerced into participation in the market on vulnerable terms. And this analysis has to be disaggregated in terms of both class and gender. The World Bank – and some members of the left – do otherwise: *Poverty and Hunger* justifies its statement that economic growth will abolish hunger and starvation on the basis of data on the *average* energy content of various countries' diets.[11] An alternative approach is exemplified by several of the chapters in this book (for example, those by Pryer and Mbilinyi) which examine vulnerability, hunger and the relation of farmers and labourers to markets in ways which look behind the averages, to the polarization and feminization of poverty, and the patterns of resistance.

Second, power and control over the terms on which markets operate are also crucial; to wield such power private individuals and companies depend centrally on the action of governments. Just as US policy has structured international food markets, and hence transformed food production and consumption in the Third World (see Friedmann's chapter in this volume), so the response of Third World states will be important, as Buttel shows, for example, in influencing the impact of Western corporations on local farming technology.

Third, the class structure of trade within the Third World is an important matter for study. The class position of traders in agricultural goods, their relations with the state, the dynamics of the markets they develop, their investments and their impact on class structure and production in agriculture – all this is beginning to emerge from the shadows. We need more work such as Barbara Harriss's in this book, which shows how diverse class structures of trade – all buttressed by state action – direct and control rural technical change and class formation.

Finally, it is important to look at ways in which some markets are abrogated rather than developed: notably labour markets. While in favour of "freeing" goods markets, the dominant ideology is generally focused on control or even abolition of labour markets. Agricultural contracting which is spreading so rapidly in the Third World as Michael Watts shows, is just one example of how new forms of unfree labour can be created amid the rhetoric of market liberalization.

Conclusion

Markets of some kind are a continuing fact of life in the Third World. This includes food markets. But "markets" and "security" are antitheses, and the World Bank has undermined its advocacy of the one by recognizing the problem of the other. If markets are extended, *in*security will increase. The left should oppose the drive to endless extension and integration of markets, and argue for bounds on market development and forms of market control for the sake of greater security. To do this, it will be necessary to prevent the development of some markets while managing others; and to change the terms on which people "come to market". The left is therefore condemned to try to understand markets better.

Notes

1. World Bank, *Poverty and Hunger* (Washington, DC, 1986a).
2. Sen, A. K., *Poverty and Famines* (Oxford: Clarendon Press, 1981).
3. World Bank, 1986a, op. cit.
4. World Bank, *Financing Adjustment with Growth in SubSaharan Africa 1986–1990* (Washington, DC, 1986b).
5. World Bank, 1986b, op. cit.
6. International Monetary Fund, "Privatisation and public enterprises", IMF Occasional Paper no.56 (1988).
7. World Bank, *Accelerated Development in SubSaharan Africa* (Washington, DC, 1981).
8. Harriss, B., "Going against the grain", *Development and Change*, vol.10 (1979).

9. Moore Lappé, F., and J. Collins, *Food First: Beyond the Myth of Scarcity* (Boston: Houghton Mifflin, 1977).
10. Elson, D., "Market socialism or socialisation of the market?", *New Left Review*, no. 172 (1989).
11. World Bank, 1986a, op. cit.

5 Food Crisis and Gender Conflict in the African Countryside

Ann Whitehead

Recent concern over Africa's food crises has focused attention on the sexual division of labour in African agriculture. One stereotyped view is that food crises have arisen because the economic changes of the twentieth century have relegated rural women to food production within an under-resourced "subsistence sector" of small-scale agriculture. This chapter contests this simplification, arguing for a more complex understanding of the interlinkage between the changing structure of gender relations in African farming households and the crisis in food production and availability.

Both men and women experience problems of food provision and impoverishment arising from external pressures, but they face these problems differently and in ways that may produce severe gender conflict. The structures of households, the sexual division of labour in farming, and misogyny in external agencies, all contribute to this. The gender conflicts which this chapter identifies are a response to economic stress and poverty among many strata of peasants, and are a symptom of economic crisis. Where crisis takes the form of gender conflict, what is happening to peasants as a whole can be masked, especially since women supporting households without male incomes can be the most severe and most politically invisible casualties. Women farmers' relation to imperialism is different from that of male farmers, and these differences must be the focus of economic and political analysis.

Are women relegated to a subsistence sector?

The early and mid 1980s in sub-Saharan Africa were marked by full-scale famine in some countries and precarious rainfall in others; also by international awareness of a deepening crisis in food security in the continent as a whole. Debate about the possible causes of, and remedies for, this food crisis coincided with the rise of a vociferous "Women in Development" lobby which pointed out the very substantial role that women play in food production in sub-Saharan Africa. As a result, discussion focused on the link between the food crisis and women's role in African food production. A dominant view developed that socio-economic change had resulted in rural women being "relegated to the subsistence sector".

This model pictures African agriculture as sharply divided between a low-productivity "subsistence" sector with unimproved techniques and a cash-cropping sector of modern high-productivity techniques. In *Women's Role in Economic Development*,[1] Ester Boserup used this model to popularize the idea that sub-Saharan Africa had initially been a female farming area, and that modernization had captured men but had left women behind. So the idea arose that African agriculture exhibited a dualism based on gender: a cash crops sector in which *men* grow highly productive income-earning export crops, and a food crop sector in which *women* use traditional methods to produce food for their families to consume.[2]

It is still not widely realized that research has shown this model to be wrong. Production data show that export crop production and food production tend to rise and fall together. Cash crops and food crops are produced by a wide variety of techniques. Food crops are also grown as cash crops. Research on the sexual division of labour has shown that Boserup overstated the extent of female labour in African farming systems and underestimated the involvement of women in the "modern" sector of the economy.[3]

The myths about a separate subsistence sector, and its feminine nature, imply that solutions to both the food crisis and women's farming problems are technical and developmental in nature. The remedies proposed centre on retaining smallholder agriculture, and improving techniques and productivity through better inputs, while protecting or bolstering women's access to resources for farming. However, what these proposed remedies ignore is the complex and often negative impact of development on rural African women, which suggests more intractable problems. The connections between women's role in food production and the changing nature of African agriculture lie, it will be argued, in the complex historical processes of commoditization and in women's specific position within those processes.

The effects of commoditization for women

Behind the stereotype just outlined is another series of stereotyped views about the historical processes by which African rural households have been incorporated into the market economy. Rural women are seen either as "left-behind wives" in semi-proletarianized rural households from which the migrant labourer husbands are absent, or as wives responsible for self-provisioning in peasant households while their husbands produce for the market. These stereotypes would only have fitted the situation in Africa in selected countries in the 1930s and 1940s, if at all. They are based on the research findings of that period. More recent research suggests greater variation and historical complexity and socio-economic change in the African countryside.

First, the effects of rural male labour migration on agricultural productivity have varied. Women's farming, or the feminization of farming as a result of the absence of men, did not necessarily imply agricultural stagnation. The effects of the loss of a proportion of male labour depended on such factors as the overall mode of economic organization, the initial population density, and the form of the sexual division of labour. Migrant wage labour promoted agricultural innovation in particular circumstances, and elsewhere women sometimes led the move to production for the market.[4]

We now also understand how varied are the profound changes in the rural economy brought about by direct production for non-local markets. As the incomes of some households increased, so a great variety of patterns of indigenous accumulation began, with varied pace and consequences. The combination of labour migration, investment in agriculture and increased commodity production brought various forms of agrarian social differentiation and class formation. Most studies of this differentiation, however, do not consider sex difference but assume the rural household to be an unproblematic unit.

Among rural women patterns of rural accumulation have produced significant polarization. Some women have benefited from increased rural incomes while others have been marginalized. Women have benefited most in the commercial farming and trading sector, where they have opportunities for increased agricultural income as well as enhanced opportunities for trade and other forms of production, such as beer brewing. Simultaneously, another group of rural households has emerged which lacks the resources to meet consumption needs. This group contains ever more female-headed households without adequate land, who join the increasing number of casually employed rural female wage workers (see Mbilinyi's chapter in this volume). This kind of women's agricultural wage labour is still largely invisible. Indeed, the stereotyped views about labour migration referred to above have hidden the very substantial amount of rural to rural migration of women associated with rural labour markets.

The implications are clear: "rural African women" do not comprise a single category of rural actors. There are important economic and other differences between them.

Changing gender relations

Most research on rural Africa focuses on socio-economic differentiation but excludes the second main change brought about by commoditization: the changing relations between men and women. There has been an enormous upheaval in domestic and other kinship-based relations between the sexes. These changes are producing considerable conflict in the African countryside and in some cases it is hardly an exaggeration to say that this amounts to a sex war. Not enough attention is being paid to the political consequences of these conflicts, or to their exacerbation by development planning and projects.

These conflicts are widespread. For example, even where female-headed households are a long-established element of kinship systems, their existence may still bespeak gender conflict, even of a serious kind. The growing number of female urban migrants provides further evidence of rural struggles. This migration can be a flight towards autonomy by rural women, fleeing from men's authority, from economic exploitation, or from lack of economic opportunity. Most important of all for this chapter is the growing evidence of struggles over new rights and obligations between household members: these are about the use of household labour, especially the labour input to cash cropping, about the sexual designation of tasks and about the distribution of household income and how it is spent.

Impressionistic evidence also suggests that a major crisis over marriage itself is occurring in rural areas. In some cases marriages are difficult to secure and do not last; in others there is intensified pressure on women to marry and to remain married, and evidence of oppression within marriage. This is reminiscent of the crisis over marriage which so exercised indigenous male élites ("our women are escaping") and the colonial male élites ("too many runaway wives") in many African colonies between 1930 and 1960.[5]

It has always been difficult to analyse these moments of

moral panic around marriage and "women". It often seems that the "crisis over marriage" is mainly discursive, an expression of male anxiety about any kind of change in the position of women. The concern with changing forms of authority and dependence is real. It seems to me to arise out of attempts, by both women and men, to make the best, as they see it, for themselves out of the complexity of real dislocations within gender relations. Intellectuals (African and "Africanist") are squeamish about addressing this area of social conflict, doubtful perhaps that gender forms the proper subject-matter of politics. In avoiding it they impoverish their understanding not only of an important area of rural social relations but also of the forms that economic pressure on peasants takes.

Women's economic roles in the farm family

Taking these gender conflicts more seriously requires careful examination of the nature of the African farm household and women's economic position within it. Most African women have always done, and still do, independent work. They are not expected to rely economically on their husbands or families but to have a separate sphere of their own work. Earlier this century, there were geographic areas where, in addition to other economic activities, women grew the bulk of food crops, most of which were consumed by their immediate and extended families. Elsewhere they did a great deal of trade and marketing, while food farming was the men's responsibility.

Women's and men's work was situated in a sexual division of labour growing out of domestic and kinship arrangements. Within these relations labour was exchanged between men and men, and between men and women. Although social exchange of labour was not confined to households, many women were members of households in which they were regarded, like the younger men, as lower in status than its senior men or its male head. These men could call upon their social inferiors to work for them. One of a wife's most significant obligations was to work for her husband and his senior close kinsmen.

The effect was that many women combined farming independently for themselves with work done as unremunerated labourers on the farms of others. This provided two very different kinds of social environment for their economic effort. In their independent work, women required effective access to resources including land to farm. Complex conventions surrounded a woman's rights to dispose of the crops she produced and her obligations to share them with her children, husband and others. The work she did for a husband and other senior men was also surrounded by complex claims and obligations. Here the return for her labour was not direct, and if conceptualized as return at all, was seen is the context of her general rights to welfare and maintenance as a household member or wife. Occasionally some work took the form of a contract between husband and wife, the terms of which could be negotiated.

All these arrangements were in the context of a domestic economy in which there was no assumption of an automatic sharing of resources in marriage. Land, cattle, hoes, money, clothes, domestic utensils and much else tended to be owned separately by husband and wife. So, too, it has been rare for there to be a joint family budget or single common purse out of which family needs are met. Rather, the separate resource streams of husbands and wives, which were the basis for their independent economic activities, also entailed ways of keeping incomes separate. These often included conventionally divided responsibilities for different aspects of household spending and consumption, for example for the clothes or medical and other needs of children, and a complex division of responsibilities for providing different items of food.[6]

The duality of women's economic role and the complex arrangements for joint and separate economies within the household are very important when we come to examine the effects of commoditization. As more land is taken into production and as it becomes scarce, so women have difficulty in protecting their land rights on the basis of either local or state codified procedures and laws. The resource base for their independent farming is undermined. Their lack of capital to purchase inputs also makes it increasingly difficult for them to pursue independent farming. At the same time an increasing proportion of a woman's labour time is spent in production for her husband, and wives' labour becomes relatively more important within the total family labour supply.

When African households' cash requirements were increased by colonial rule either directly (by tax demands) or indirectly (by new consumption goods), the main immediate avenues for earning such income was men's cash cropping or migrant labour. Women members of the household were able to make their contribution to increased cash needs by their work as family labour in cash cropping or by increased trading. In the initial phases of these processes, in so far as women's welfare was bound up with that of their households, there were simple incentives for them to do so.

However, over time these decisions to undertake more work as "family labourers" took on new economic meanings. As the terms of trade declined against peasants, as land became scarce, and as rural differentiation proceeded, there was increasing evidence of acute stresses and strains. As Pepe Roberts explains, the wives' situation as "unfree labour became increasingly important to the household as commodity relations destroyed other bonds securing non-free labour (e.g. that of sons) to the peasant household".[7] This is reflected in several reports of increased rates of polygyny as well as of conjugal conflict as commodity production increased. There is no guarantee that the increased labour time now required for peasant reproduction bears equally upon men and women, and the potentiality for coercion within the customary obligations of a married woman to her husband may become an important element in her increased work load. Among the poorer peasantry female-headed households emerge as a response to this economic stress.

As the rural division of labour becomes more firmly based on marriage and the (often smaller) household, stress and conflict are also apparent over the control of resources. Conventions governing the distribution of rewards in kind within the household (which may or may not have been equitable in the first place) do not automatically lead to sharing when household product is increasingly in the form of money, and especially when, for a variety of reasons, it goes to men only. Hence critical differences of opinion arise between men and women as to how household income should be spent in the interests of household welfare. Some authors associate this with lower nutritional well-being. Again many studies report women complaining of not being able to command access to money income which appears as belonging to the husband when produce is sold from "his farms", because of the lack of any concept of common income and joint resources within marriage.

The role of development planning and projects

Whether or not they are successful in their own terms, many rural development projects funded by a combination of overseas and national agencies appear to make women worse off. Small wonder then that there are often reports of women resisting these changes; in some cases projects have failed because of women's resistance to their role within them. Examining these projects, we find they have a depressingly similar format. They ignore the scale and significance of women's independent farming or income generation, leaving it unmodernized or making it impossible; simultaneously, they recruit women as family labour to their husbands' fields.

The historic neglect of food crops in agricultural research is one reason for this. The data base is a second problem. Women's role in farming is partly ignored or misunderstood because their work was, and remains, largely invisible in national statistics. This results partly from international conventions about what constitutes work (mainly production for the market, or paid work); from data collection methods; and from stereotypes that both African and European men hold about women's contribution to the economy. But a third, and least widely recognized, problem of development planning arises simply from widespread, systematic sexual discrimination against women in agricultural delivery systems.

I say "simply" from sexual discrimination because not only are there few women in agricultural extension work, and not only are women's crops not targeted for improvement, but innovative, efficient and resourced women farmers are ignored in favour of less endowed and less efficient male farmers.[8] A major source of these attitudes is in Western development and agricultural extension training.

It is a major failing of development projects that they are based on a conceptual model of the African family farm which does not reflect the complex and particular forms of their social relations described above.

The planners' model centres on the idea of the conjugally based household as an economic enterprise in which the members work together. The husband/father is regarded as managing the resources on behalf of other members, and those others, conceptualized as his dependants, provide labour under his direction. Hence the projects require a family labour input from household members (especially wives) other than the putative male head. Sub-Saharan African domestic organization, as shown, is emphatically not of this kind.

The difference between the planners' model and rural reality has important consequences. These are illustrated in a well-known example – that of largely unsuccessful attempts to introduce irrigated rice production in the Gambia.[9] The lack of success stemmed in part from just this "male-dominant", "domestic sharing", model of the household which shaped the project. An initial assumption was that the men were rice growers with full control over the necessary resources. Incentive packages included cheap credits, inputs and assured markets offered to male farmers. But it was women who traditionally grew rice for household consumption and exchange, within the kind of complex set of rights and obligations between husbands and wives discussed above. The scheme proposed to develop irrigated rice production on common lands to which women had secured use rights. Backed by project officials, men established exclusive rights to these common lands, pushing the women out to inferior scattered plots to continue cultivating traditional rice varieties. All access to inputs, labour and finance was mediated through husbands, and women became notably reluctant to participate in their planned role as family labour. Husbands had to pay their wives for what work the women did do on the irrigated rice fields. Dey and others showed that the disappointingly low levels of improved rice production arose substantially from these misunderstandings.

The creation of dependence

Whether considered as part of the general changes associated with commoditization or in the more acute form associated with development projects and planning, the trends outlined represent a highly significant historical process. For African women in peasant households, recruitment primarily as family labourers represents the construction of a hitherto rare form of dependence within and on marriage. This conjugal dependence implies that women's economic opportunities are not enough for their independent survival. A woman can increase overall family welfare by working on her husband's farm or enterprise, but in conditions where she rarely has guaranteed access to household income and welfare. As a woman's independent capacity for survival diminishes, she finds she lacks leverage over resources within marriage.

The increase of conjugal dependence was one of the important changes brought to marriage by industrialization and proletarianization in nineteenth-century Europe. Over time women became economically much more reliant on the waged work of their husbands. This was as a result of their differently developing relation to the labour market, and because of the development of domesticated reproduction in a particular form. Feminist historians suggest that women's loss of autonomy was not experienced without considerable struggle and upheaval. What is happening in Africa is problematic in the same general way, although it takes a different form, both because of the different nature of family arrangements in Africa, and because of the differences in the socio-economic change experienced.

Women farmers' dilemmas

African women farmers may be placed in difficult and contradictory situations in which an unenviable choice may have to be made. They may, for example, have to choose between loss of autonomy and poverty. A woman's autonomous sphere lies in her independent income-earning opportunities in farming, but these are diminishing (though those in trade may remain) and provide relatively low incomes and few opportunities. Wages and conditions for casual wage labour, the main independent earning alternative, are very poor indeed. As a "family labourer", a woman may produce more crops and more income: as an unpaid labourer for her husband she may become better off if she helps him become successful. More important, she may also feel that her children's welfare is more secure. However, in addition to a lack of control over spending and welfare decisions, as unpaid workers women do not build up their long-term resources. The decision to do more unpaid family work may hook a woman into a dependence which leaves her very insecure at times of crisis. The growing number of poor female-headed households and the crisis over marriage in Africa discussed earlier are evidence for this. Mbilinyi's chapter in this volume offers just one chance to hear rural women's voices on this subject.

The divided family

Misogyny in development and planning agencies combines with a disadvantageous resource-holding structure to produce considerable divisions between rural men and women. The reluctance to address this social conflict arises partly because of ideologies which protect family and domestic behaviour from public scrutiny, but it also looks uncannily like a male alliance between "patriarchal" male farmers and sexist male bureaucrats. However, such an alliance is not formed simply because of the existence of the biological categories of man and woman. To understand it we need to look beneath the form of the immediate gender conflict to its underlying content. In sub-Saharan Africa this content is the central importance of the sexual division of labour to peasant production and the particular character of the social relations between men and women within peasant households. Economic change and associated food crises have been accompanied by intensified gender conflict. The sex war in rural areas is a response to economic stress and poverty, and it takes its form and seriousness from the deteriorating economic conditions faced by a majority of peasant producers. Experiencing this deterioration on the terrain of sexual politics may obscure the fact and politics of widespread impoverishment.

Consider, for example, the growing number of female-headed households. Abandoned by their husbands and sons, women blame modern men's low standards of personal responsibility, while men argue that women drive them away with unacceptable sexual or domestic conduct and their new desires for personal freedom. The mutually expressed anger and disappointment obscure poverty as a source of crisis. The contradictions set up in the farming system produce impossible economic conditions for peasant households but men and women experience these contradictions very differently. It is in this light that we should reconsider the struggle over household income described earlier. Rural women complain at the selfish way men spend; men complain reciprocally at the constant financial demands from wives. In reality husband and wife have conventionally been allotted different responsibilities for spending, which come into conflict in conditions of economic stress. Women may be particularly responsible for short-term spending – for example, for food – while men are responsible for long-term spending, especially the purchase of farming inputs. Both are essential to survival but where there is not enough income to meet both, impoverishment is experienced as sharp domestic conflict.

This division of responsibility within the farm family has implications at much wider levels. Women's increased dependence within marriage has been identified in this chapter as a major change in rural gender relations and as a source of gender conflict. Women farmers' relation to imperialism is different from that of male farmers; the differences must be the focus of economic and political analysis, not only for reasons of equity but because crisis in the form of gender conflict can mask a more general crisis of the peasantry.

Notes

1. Boserup, Ester, *Woman's Role in Economic Development* (London: George Allen and Unwin, 1970; new edition, Earthscan Publications, 1989).
2. Discussed further in Ann Whitehead, "Women in rural food production in sub-Saharan Africa: some implications for food strategies", a paper given to a Symposium on Food Strategies at the World Institute for Development Economics Research, Helsinki, July 1986, funded by the United Nations University. I am grateful to Maureen Mackintosh for her initial comments on that paper, and for her editorial help in preparing this article from it.
3. See, for example, Paul Richards, "Ecological change and the politics of African land use", *African Studies Review*, vol. 26, no.

2 (1983); and Jane Guyer, "Women in African rural economies: contemporary variations", in J. Hay and S. Stichter (eds), *African Women South of the Sahara* (London: Longman, 1984).

4. See, for example, Gavin Kitching, *Class and Economic Change in Kenya* (Princeton: Yale University Press, 1980); and Sharon Stichter, *Migrant Labourers* (Cambridge: Cambridge University Press, 1985).
5. Megan Vaughan and Ann Whitehead, "Overview paper: the crisis over marriage in colonial Africa", for Workshop on the Crisis over Marriage in Colonial Africa, Nuffield College, Oxford, December 1988.
6. See Ann Whitehead, "I'm hungry mum: the politics of domestic budgeting", in K. Young, R. McCullagh and C. Wolkowitz (eds) *Of Marriage and the Market*, (London: CSE Books, 1981).
7. Pepe Roberts, "Feminism in Africa; feminism and Africa", *Review of African Political Economy*, no. 27/28 (1983).
8. K. Staudt, "Agricultural policy implementation: a case study from Western Kenya", in *Women's Roles and Gender Differences in Development*, Case Studies for Planners prepared by the Population Council (West Hartford: Kumarian Press, 1985).
9. J. Dey, "Development planning in The Gambia: the gap between planners' and farmers' perceptions, expectations and objectives", *World Development*, vol. 10, no. 5 (1982); see also Judith Carney, "Struggles over crop rights and labour within contract farming households in a Gambian irrigated rice project", *Journal of Peasant Studies*, vol. 15, no. 3 (April 1988).

6 Taking the Part of Peasants?

Henry Bernstein

Taking the part of peasants, or of family farmers, is the slogan of an agrarian populism which has deep historical roots, and which continues to attract the sympathies of many socialists. The aim of this brief chapter is, from a socialist position, to cast a critical eye on the character of peasants in contemporary capitalism (imperialism), to argue against viewing peasants as a single social category or (exploited) "class", and to interrogate what it means to "take the part of peasants".[1]

Of virtuous peasants and vicious states

Peasants are often viewed as the representative victims of the modern world while, at the same time, widely held images depict them variously as expert ecologists and agricultural innovators, exemplars of a benevolent small-scale production, militants of the "moral economy", the driving force of liberation struggles. In different ways these images resonate deep ideological opposition to the generalization of commodity production (i.e. capitalism), to the brutalities of both its historical formation and its everyday reproduction in the contemporary Third World.

This opposition frequently draws on emotive if idealized conceptions of *family* and *community*. The values attributed to family farming (self-reliance and independence through the use of household resources and labour) and to rural community (co-operation, reciprocity, solidarity) inform one kind of critique of the possessive individualism of the market and the inevitable divisions and inequalities of capitalism. The appeal of such virtuous constructions of peasant life is reinforced by another pervasive motif of resistance to urban industrial civilization and its discontents, that of a direct relation with nature in securing a livelihood from the land.

Setting aside nostalgia for a golden past, a world we have lost, it is also argued that peasants can adapt to modern conditions and make major contributions to economic and social progress, even though they are generally prevented from doing so. This brings us more directly into the terrain of contemporary political economy and discourses of Third World "development". Peasants are held back by economic exploitation: the terms on which they obtain land and credit, sell their crops, and buy means of production and consumption; also by political and ideological oppression: peasants are dispossessed of any effective political voice and influence, their farming skills and ways of life denigrated in stereotypes of rural "backwardness" held by those who concentrate economic, political and cultural power. In particular these negative stereotypes are intrinsic to the practices of "developmental" states whose vision of modernization is urban-industrial, justifying the exploitation and oppression of peasants that the state organizes or otherwise colludes in.

In the opposition of peasant virtue to the viciousness of states, current emphasis on the latter is so powerful because it provides a point of convergence for otherwise diverse ideological currents. For radicals, the colonial state was as central to the initial and forcible integration of peasants into the international capitalist economy as contemporary Third World states of different political complexions are to squeezing peasants in the name of industrial accumulation. For free-market ideologues (like the World Bank/IMF and its organic intellectuals – see Mackintosh's chapter in this volume), state economic activity – or "interference" with the market – seeks to reap a bureaucratic rent from the exchange of commodities, distorts prices and incentives, and thereby stifles the energies of the small farmer or peasant as economic man (*sic*).

Anti-statism, as we know only too well, is a dominant ideological motif of the reactionary offensive in the current period, however selective and opportunistic its uses in practice. It has a particular force and fascination – and ability to confuse – in this context because anti-statism is also a banner of agrarian populists who come from the left and "take the part of peasants" equally against capitalist, state capitalist and state socialist regimes.

Socialists cannot afford to be dismissive of the criticisms and claims of agrarian populists. Indeed many of their best points derive from the socialist tradition, for example:

- the centrality of the state to capitalist development (with due regard to its wide range of historical variation)
- the characteristic divisions of labour of capitalism that also divide those whom it exploits and oppresses, including the divisions of industry and agriculture and of town and countryside, to the detriment of the latter
- the different ways capital enters agricultural as distinct from industrial production, and the reasons (see Watts's chapter in this volume)
- not least, the issues that the distinctive features of agriculture and the "peasant question" in capitalism pose for agrarian strategy in transitions to socialism, manifested in unabated controversies about the experiences of the Soviet Union, China, Cuba, Vietnam or Mozambique.

However, what we have to ask of the populist position is this: are peasants uniformly so virtuous (and embattled), states necessarily so vicious (and powerful)? Or is reality more contradictory and complex than the essential dualism of agrarian populism – peasant family/community vs. state/capital – suggests?

Peasants in capitalism

First, we should dispense with any notions of peasants as "precapitalist" or only partially incorporated within capitalism, whether expressed in theories of the "articulation" of modes of production, in so-called orthodox Marxist views that peasants will – and should – inevitably disappear under capitalism, or in romantic views of the ability of the rural "moral economy" to resist commoditization.[2]

The starting-point must be to view peasants today as agrarian *petty commodity producers within capitalism*. This is also the position of more sophisticated populists, although I would argue that they do not theorize it adequately or follow through its implications. Petty commodity producers are, in a crucial sense, both capitalists and workers at the same time: they own or have access to means of production which they "put to work" with their own labour. As capitalists, they employ – and therefore exploit – themselves.

(Self-)exploitation is not necessarily distributed evenly within peasant households, however. Commonly, various (and changing) forms of patriarchy structure the appropriation of the labour of women and children by male household heads and other senior men. Thus gender relations are one major source of divisions, and possible conflict, within peasant households and communities (see Whitehead and Mbilinyi in this volume).

Class differentiation of the peasantry and its dynamics were emphasized by Lenin and Mao Zedong, while populism tends to champion "middle peasants" as representing the "natural" condition of the peasantry, "the backbone of peasant society".[3] However, the formation of relatively stable middle peasantries in contemporary capitalism needs to be explained no less than the erosion of middle peasants through differentiation. In both cases, explanation requires investigating the specific historical circumstances of such peasantries – their relations with other peasant classes, with rural labour, with markets, with forms of agribusiness and, of course, with the state.

On one side of the middle peasants are poorer peasants, many of whom are unable to reproduce their means of production in the face of multiple pressures – including competition with other peasants over land, labour, access to inputs, to credit or to markets. This means that they become marginalized as farmers or are ultimately dispossessed and proletarianized. On the other side, rich peasants are those who have been able to accumulate and to employ the labour of others.

None of this means that class formation among the peasantry happens everywhere, in the same way, or with the same results, as mechanical applications of Lenin's schema of differentiation assume. The extent, character and stability of class formation among the peasantry are always the outcome of specific historical processes of competition and struggle between peasants, and between peasants and others. Often the latter obscures the former, with processes of rural class formation masked by solidary ideologies of community, "tradition", custom, etc., generated as a means of defence against the depredations of "outsiders": land grabbers, merchants, agribusiness companies or the state itself.

This brief sketch of differentiation indicates that the most widespread manifestations of the contradictions of capitalism – the divisions, exploitation and oppression of class and gender relations – are as much part of peasant economy and life as of other spheres of productive and social activity in capitalist societies. They are *internal* to the peasantry, generated by its relations of production (the combination of capital and labour in petty commodity production), as against the populist view that differentiation is "exceptional" and/or "*external*" to essentially egalitarian peasant communities, the divisive result, for example, of state patronage which privileges some peasants against others.

Populists view the relations between peasant communities and the "external" world of wider divisions of labour, markets and state structures primarily as relations of exchange (though often reinforced by political domination). Hence peasants are "exploited" through the terms of trade for the commodities they sell and buy in markets controlled by others, in Africa above all by the state. Presumably, then, should the terms of trade improve for peasant producers, their "exploitation" is commensurately reduced; should the conditions of exchange attain a sufficiently virtuous equilibrium, peasants would no longer be an exploited "class" at all.

The strategic differences between populist and socialist perspectives on "the peasant question" are not only theoretical, as sketched above, but also practical and political. This becomes evident if we examine the two main prescriptions of agrarian populism in the African context: the freeing of peasant commodity exchange from state control of marketing, and of peasant farming from "development" schemes, imposed in the name of "modernization" by unholy alliances of state, foreign aid and international capital, which so often prove to be technically inappropriate, economically irrational and politically repressive.

The first prescription resonates a version of "people's capitalism" for peasants, enabling them via more competitive markets to get the best returns from controlling their own means of production and managing their own resources (a familiar petty-bourgeois idyll). This holds out the prospect, at least in theory, that through the correct reforms the emancipation of the peasantry can be realized *within* capitalism. So there is no *necessary* contradiction between peasants and capital, or the state (as there is between workers and capital) after all.[4] Peasants cannot, then, constitute an intrinsically "exploited" class in capitalism if the condition of their freedom is the "freedom of the market".

The dialectics of commoditization are such that market liberalization may well stimulate some initial increase in peasant commodity production in parts of Africa, and hence contribute to agricultural growth for a period, while at the same time the process is likely to accelerate tendencies of differentiation that already exist, to enhance opportunities for accumulation by some peasants and to increase the impoverishment and insecurity of many others – precisely the effect of the "normal" working of the market, as Maureen Mackintosh points out in this volume.

The first prescription thus contains the seeds of the same contradiction that Chris Scott analysed in the related case of "neo-classical populism":

> The desire for the simultaneous achievement of efficiency and equity is the contradictory core of neo-classical populism. Perfect markets for commodity exchange are advocated for inputs and outputs, to be sustained on a rural social foundation consisting of communities of egalitarian peasant freeholders. . . .This manifesto for the peasantry with its populist attacks against rapacious landlords, city "folk" and an oppressive plutocratic State calls for efficiency without inequality, statics without dynamics and capitalism without exploitation – in short, for the lion to lie down with the lamb.[5]

The second prescription is significant more for what it omits than what it says. Much populist criticism of the ideology, mechanisms and outcomes – farcical or disastrous – of state-promoted rural "development" projects is shared by a wide spectrum of socialist and liberal opinion. What it omits, however, is any consideration of the place of agriculture and of peasants in a comprehensive agenda of progressive change at the national level. The principal thrust of agrarian populism is indeed a laissez-faire one: "hands off the peasants – they will be fine if left in peace to manage their own resources in properly competitive markets that reward their efforts".

Peasants and socialism

Understanding the complexities and contradictions of "actually existing capitalism" is a necessary first step in considering socialism, as Maureen Mackintosh (in this volume) shows in relation to the analysis of markets. Farming in capitalism tends to vary much more – by its forms of social organization (different types of household labour, wage labour, contract labour), range of size and scale, degrees of capitalization and mechanization – than the enterprises to which it links backwards (input manufacture) and forwards (processing, marketing), where agribusiness capital concentrates and whence it seeks to control agricultural production (see Watts's chapter in this volume). This means that peasant farming is no less "typical" of capitalism than, say, plantation agriculture.

While the historical record of the oppression of peasants in the name of "development" and "modernization" is not as uniform as populists suggest (and regimes that oppress peasants tend to oppress workers as well), much of it lends appeal to the populist case, especially when oppression is exercised by self-styled socialist regimes. However, the key issue remains investigating alternatives both to the blank cheque populism writes for peasant virtue and to the model of industrialized agriculture dear to bureaucratic "socialism". Both embody misconceptions about agriculture in capitalism (and by extension about capitalism more generally).

A "people's capitalism" for peasants is utopian, and privileges peasants above those who have no access to means of production – workers and the dispossessed and marginalized, both rural and urban (see Pryer's chapter in this volume). Bureaucratic socialism is mistaken in its view that large-scale industrialized farming is (a) the "essential" form of agricultural production in capitalism, (b) necessarily the most "efficient", and (c) that its "superior" forces of production (and forms of organization) provide the model for socialist agriculture to appropriate or emulate.

As peasants are not a uniform social category or "class", we have to ask: taking the part of *which* peasants? in what circumstances? for what reasons? by what means? This is more demanding of both theory and practice than awarding the "freedom of the market" to all peasants on the assumption that they share the same conditions of existence, aspirations, goals and interests.

Asking "which peasants?" is necessary to identify those who, due to their social position (in class and gender relations, divisions of labour, particular ecological conditions, systems of exchange), would benefit from socialist co-operation, in however preliminary a form to start with, and whose participation can be encouraged by *immediate benefits* in terms of the enhanced access to resources, control over production, and security of subsistence that co-operation makes possible.

Establishing and developing socialist co-operation from peasant farming, and resolving the contradictions it inevitably confronts, is first and foremost a *political* project of changing existing social relations and practices. It cannot be technocratically determined by abstract criteria of "efficiency" or by bureaucratically imposed "modernization from above".[6] The question of "*which* peasants?", of the allies of socialism in the countryside and their role in transition, is thus inextricably linked with asking: what kinds of political process, methods of work, forms of organization, types of state structure and practice can enable and guide social transformation, can overcome the divisions of industry and agriculture and town and countryside, in the interests of *all* those exploited and oppressed under capitalism?

The strategic questions indicated are open-ended; they do not admit to dogmatic answers as we know from the history of struggles for socialist transformation, always intensely contradictory and often bitter, in a world dominated by imperialism. Socialists have rich resources of historical experience and theory to tackle such difficult issues, without succumbing to the blanket advocacy of "the peasant way" declared by agrarian populism (or the vision of agricultural "modernization" held by bureaucratic socialism).[7]

It is time to jettison the morally charged dualism of peasant virtue and state viciousness, and the theoretical conceptions called into its service by agrarian populism. Moving beyond "actually existing capitalism" requires transformation of the peasantries as well as of the state regimes that it contains. The multiple contradictions of their social existence within capitalism means that peasants themselves, in alliance with the political struggles of others, may play a central role in advancing or hindering this revolutionary process.

Notes

1. The title is adapted from Gavin Williams, "Taking the part of peasants: rural development in Nigeria and Tanzania", in Peter Gutkind and Immanuel Wallerstein (eds), *The Political Economy of Contemporary Africa* (Beverley Hills: Sage Publications, 1976) (2nd edn, 1985), a representative statement of agrarian populism in the African context. The perspective taken here owes a great deal to Peter Gibbon and Michael Neocosmos, "Some problems in the political economy of 'African socialism'", in Henry Bernstein and Bonnie K. Campbell (eds), *Contradictions of Accumulation in Africa. Studies in Economy and State* (Beverley Hills: Sage Publications, 1985), and Richard Levin and Michael Neocosmos, "The agrarian question and class contradictions in South Africa: some theoretical considerations", *Journal of Peasant Studies*, vol. 16, no. 2 (January 1989).
2. The central idea of the articulation of modes of production is that capitalism "articulates" or combines with pre- or non-capitalist modes of production which it simultaneously "conserves" and "dissolves" in line with its needs for cheap labour power (rurally based labour migration systems) and cheap commodities (produced by peasants). The best source in English on this theoretical approach and its applications remains Harold Wolpe (ed.), *The Articulation of Modes of Production* (London: Routledge, 1980), including Wolpe's substantial introduction. The best source for the application of the "moral economy" concept to peasant societies is the original one – James C. Scott, *The Moral Economy of the Peasant: Rebellion and Subsistence in Southeast Asia* (Princeton: Yale University Press, 1976).
3. Williams, op. cit.
4. In fact the universal political weakness of peasants is as misleading a generalization as their universal economic "exploitation"; see

Patnaik in this volume on rich peasant power in India. While the "peasant interest" (typically articulated and organized by rich peasants on their own behalf) is not represented at the level of the national state in most African countries, it often exercises considerable influence over local state structures. Williams himself provides an instructive example, with apparent approval, of how rich peasant-traders were able to gain control of resources provided by a World Bank project in northern Nigeria: Paul Clough and Gavin Williams, "Decoding Berg: the World Bank in rural Northern Nigeria", in Michael Watts (ed.), *State, Oil and Agriculture in Nigeria* (Berkeley: University of California, Institute of International Studies, 1987).

5. Scott, Chris, "Review of Keith Griffin, *The Political Economy of Agrarian Change*", *Journal of Peasant Studies*, vol. 4, no. 2 (January 1977).
6. A lucid summary of the advantages of co-operation in peasant farming without any changes in existing (hand tool) technology is given by Michaela von Freyhold in *Ujamaa Villages in Tanzania: Analysis of a Social Experiment* (London: Heinemann, 1979), pp. 22–5. They include economies of scale through the combination of labour, pooling knowledge and facilitating planning, complementation effects and timing effects, all of which help establish social conditions for the more effective introduction of new technologies subsequently.
7. Ashwani Saith (ed.), *The Agrarian Question in Socialist Transition* (London: Frank Cass, 1985), is a valuable collection drawing on these resources; see also the articles on the experience of Nicaragua in the *Journal of Peasant Studies* by David Kaimowitz (vol. 14, no. 1 (1986)), Michael Zalkin (vol. 16, no. 4, 1989) and Max Spoor (vol. 17, 1990).

7 Some Economic and Political Consequences of the Green Revolution in India

Utsa Patnaik

The inception of India's "new agricultural strategy" from 1961 followed the visit of a team of experts from the Ford Foundation, whose report on "India's Food Crisis" evidently greatly influenced government policies. Under the new strategy the Intensive Agricultural Development Programme (IADP) was started in fifteen districts selected for the availability of irrigation and absence of acute tenurial problems. High-yielding cereal seed varieties already developed in Mexico and Taiwan, along with fertilizers, pesticides and credit to farmers, were made available at subsidized rates. The high-yielding varieties programme was subsequently extended to all 324 districts in the country.

Government policy documents of that time bear the stamp of ideas about "modernizing" agriculture such as those of T.W. Schultz.[1] Policy documents stressed the need to "transform traditional agriculture" by introducing an entirely new, imported technology package, while the need to reform land-holding structure and tenurial relations receded to the background.

Two phases of the Green Revolution: the uneven expansion of output

The first phase of the Green Revolution from the early 1960s to the mid 1970s was primarily concentrated on wheat and was associated with a substantial rise in both yield per unit area and total output, especially in North India. In this phase foodgrains prices were rising faster than prices of either manufactured inputs into agricultural production, or final consumption goods. Thus, it took less wheat output, for example, to buy fertilizers or fuels in 1974 than it had before the start of the Green Revolution in 1963. The barter terms of trade moved sharply in favour of the agricultural sector. At the same time, real earnings of rural labour were declining. The profitability of producing cereals rose significantly, especially in wheat where the new varieties gave a substantial yield gain under field conditions. Considerable capitalist investment was visible, especially in the wheat-growing region of North India (comprising Punjab, Haryana and Uttar Pradesh), in the form of irrigation pumpsets, threshers and tractors in addition to higher working capital outlays on the new inputs and on wages. On the other hand the relative profitability of pulses, that part of foodgrains which provides the main source of protein for the rural poor, fell and a declining trend of pulse production per capita set in.

In the primarily rice-growing coastal and river delta areas of high rainfall and canal irrigation, the new varieties were found to be less successful in terms of yield rise. Problems of water management and high atmospheric humidity during the long periods of cloud cover raised costs and limited yields in the field by promoting pest infestation. Much less capitalist investment took place than in wheat-growing regions, and widespread petty tenancy continued. In North India, however, there were several rounds of tenant evictions by landowners switching to direct cultivation with hired labour, and the proportion of hired labour in the rural working population rose sharply.

During the second phase of Green Revolution, from about 1975 to the present, the Punjab experience has not been replicated, whereas the advocates of the new strategy anticipated that the new techniques would be generalized wholesale to rice-producing areas. This has not happened. Instead, high-yielding rice has emerged as a second crop grown primarily for sale in the traditionally wheat-growing region of North India which had already benefited the most from the first round of technical change. Over 90 per cent of the Punjab rice crop today is sold, compared to barely one-third to two-fifths in the traditional rice-growing areas where it is also the staple foodgrain of the producers. Since 1975, the rate of increase in foodgrain prices compared to manufactured goods prices has slowed down.[2] In non-foodgrains of commercial importance, such as cotton and sugar cane, the price rise even during the first phase had been much less than in foodgrains; the recent unfavourable shift in the terms of trade has hit producers of these crops particularly hard. Waves of peasant agitation for higher crop prices have erupted during the last decade.

National self-sufficiency combined with mass poverty

The new strategy has been in operation for nearly three decades; it is clear by now that the unevenness of growth has been accentuated at multiple levels. Gains have been concentrated in particular crops and regions; those gains have not been spread evenly through society but are concentrated on particular social groups. Total foodgrains output has registered an impressive expansion, trebling from 50 million tons in 1951 to 150 million tons by 1983–4 (however, it fell below this level in the following years with a sharp dip in the drought year of 1987–8). In terms of production and availability per head of population, however, the picture is disappointing in that the well-documented decline in grain production during the half-century preceding independence has not been made up. While the average Indian consumed 200 kg of grains on the eve of World War I, this had dropped to 150 kg by independence.[3] In the subsequent thirty-five years per capita foodgrains availability rose by less than 10 per cent to around 166 kg. These levels are inadequate to meet minimum nutritional requirements when there are large income disparities. A more populous and initially equally poor neighbour, China, started with a higher per capita availability in 1951 and by 1984 had raised it still further to achieve a 50 per cent higher level (about 250 kg, using comparable concepts) than India. Rural income distribution is known to be considerably less unequal in China than in India. Further, the nutritional balance of the average Indian's food intake has worsened, with a fall in the share of pulses relative to cereals.

For a quarter-century after the inception of planned development, from 1951 to 1975, India produced less foodgrains than were consumed, the balance being imported (mostly from the USA). In the decade 1975 to 1984, however, domestic production exceeded consumption, the difference going into building up food stocks and a small volume of exports. India thus became "self-sufficient" in foodgrains during this decade; indeed, the government's buffer stocks had mounted to an unprecedented level of nearly 30 million tons by 1984–5, which is 20 per cent of the peak annual output.

The concept of "self-sufficiency", defined by zero food imports or positive exports, is, however, a tricky one: it need not mean that people are not going hungry. "Self-sufficiency" in a market economy has to be understood in relation to effective demand (see Mackintosh in Chapter 4). It is quite possible for foodgrains to be exported while famine prevails, as happened during the great Bengal famine of 1943–4. At that time, nearly 4 million of the rural poor died as a consequence of the British colonial government's policy of printing money to finance its war expenditure. This policy sent rice prices soaring beyond the buying capacity of the rural masses. In contemporary India there is no famine, but there is widespread and persistent unemployment and poverty, especially in rural areas. This has led many observers to conclude that the excess-supply situation which enabled the government to build up large food stocks by 1984, even with stagnant and inadequate levels of per head consumption, was the outcome of the lack of purchasing power of the undernourished, combined with burgeoning marketable surpluses in the hands of the minority of capitalist farmers.

Macroeconomic indicators relating to incomes and employment tend to support this conclusion. There has been a process of profit-inflation in the economy. The official National Accounts Statistics show that total property income, that is profits, rent and interest, as a proportion of total income generated by agriculture and mineral extraction, rose steadily from 8.7 per cent in 1978–9 to 12.5 per cent by 1984–5. On the other hand, the real value of the rural wage bill has remained level during this period, even though the number of rural employees has gone up by nearly 20 per cent. A constant level of real income shared among larger numbers of wage workers implies declining real earnings per head. Analysis of the variation between states shows that the largest declines are in the relatively stagnant regions of Western, Central and Southern India. Labourers and poor peasants make up some three-fifths of the rural population; the contraction or stagnation of their income restrains their food consumption even though their food intake is nutritionally inadequate; the obverse of this fall is the rise in the real incomes of the propertied minority, whose food consumption is already at satiation levels, and who therefore spend a much lower proportion of their higher incomes on food. This shows at the aggregate level (given the substantial share of this minority in total expenditure) as a small decline in the proportion of total consumption expenditure going on food, or an Engel effect incorrectly interpreted by some observers as indicating a rise in the general standard of life.[4]

The Green Revolution strategy has implied a standardization of cereal crop varieties and a narrowing of the earlier traditionally diversified genetic base of crop types. Combined with the dependence of these new dwarf varieties on irrigation, this has contributed to an increase in output fluctuations compared to the period before the mid 1960s. If the rains are good and irrigation water available at the right time, yields from the dwarf varieties are high, but either drought or water-logging makes them plummet. In traditional practice, by contrast, some insurance was provided by sowing part of the areas to lower-yield but drought-resistant varieties and part to tall, flood-resistant varieties. Second, the proportion of cultivated area under effective irrigation is just over one-quarter, with rainfed and dryland agriculture relatively neglected: over one hundred districts or nearly a third of the total area is now declared to be drought-prone. The reliance placed by the new strategy on giant state-financed irrigation projects on the one hand and on private tubewell-pumpset irrigation on the other, has also led to a sharp decline in traditional small- to medium-scale water conservation systems maintained in the past by community labour. (These systems can still be seen in operation in the few remote parts of the country relatively little ravaged by the inroads of capitalism.) These problems have been compounded by rapid deforestation contributing to soil erosion and degradation as well as a lowering of the sub-soil water table.

The regional and class concentration of growth

India can be divided into three groups of states: a fast-growing region in the north (Punjab, Haryana and Uttar Pradesh), a stagnating region in the west and south (Gujarat, Rajasthan, Maharashtra, Andhra Pradesh, Karnataka, Kerala and Tamil Nadu), and the rest of India. These groupings are very rough and there are many high-growth irrigated sub-regions in the second category just as there are some low-growth areas in the first. Between 1974–5 and 1985–6, the second phase of the Green Revolution, the share of North India in total food output rose steadily from 29 to 36 per cent; the share of West and South India declined from 37 to 32 per cent; while the share of the rest of India changed marginally from 34 to 32 per cent, with some fluctuation. Thus the region of North India, which had initially the lowest share of the three in total output, accounts for the highest share today; while a far vaster region in terms of area and population, which initially contributed the highest share of total food output, today accounts for the lowest share. (Inclusion of the data for the last two years, which has seen drought affecting mainly West India and South India, would reinforce these trends.)

While the region of stagnating production has registered a mere 16 per cent rise in food output in the 1983–4 to 1985–6 period compared to the period 1974–5 to 1976–7, the region of dynamic growth by contrast has registered a 67 per cent rise in output. Population has grown everywhere by about one-fifth, so it is clear that per capita food output has been falling in most of the country while rising fast in North India. A comparison of the all-India sample surveys on employment relating to 1972–3, with later rounds in 1977–8 and 1983, shows a rise of unemployment in rural areas.

No less striking is the class concentration of gains in the areas experiencing the Green Revolution. A cross-sectional study of farms in Haryana indicates that investment in new techniques was strongly concentrated in the larger-scale, labour-hiring farms while the majority of families, who were self-employed or hired out their labour, failed to reach the poverty line.[5] A comparison of the agricultural censuses of 1981 and 1971 in the case of Punjab shows that the number of smallholdings below 2 ha fell, owing mainly to the cessation of some 0.8 million petty tenancies; land concentration increased and the average holding size went up. At the same time Punjab has registered a steep rise in the proportion of labourers in the agricultural workforce (from 32 to 38 per cent), completing the classic scenario of capitalist development in agriculture leading to the displacement of small producers and to proletarianization.

The importance of food stocks procured and operated by the government becomes evident. On the one hand, large procurement is possible because the well-to-do surplus farmers in the fast-growing regions sell almost their entire crop; on the other hand, the distribution of food at subsidized rates protects the non-rural consumer against inflation, and provides the rural landless and land-poor employment – albeit on a limited scale – under various food-for-work programmes. Nevertheless, the real incomes of the rural poor are declining alarmingly in the low- and negative-growth regions.

The Green Revolution region of North India not only accounts for one-third of total foodgrains production: it contributes 97 per cent of the government's procurement of wheat and 65 per cent of its total rice procurement. Within the region, the state of Punjab alone accounts for respectively 61 per cent and 44 per cent of total wheat and rice purchased by the government. It is this granary of India which has seen, in the last few years, an unprecedentedly violent movement, albeit supported only by a minority, for secession from the Indian union. The agrarian origins of this movement call for a brief discussion.

Increasing national divisions: Punjab and Assam

Extreme disparities of growth by region and class give rise to special problems in a nation which is a union of nationalities. Numerous cultural and social contradictions, which were suppressed and transcended in the course of a common anti-imperialist struggle before independence, re-emerge and acquire an ominous strength as the post-independence path of capitalist industrialization runs into difficulties. The very foundations of a nation based on a union of nationalities can be threatened primarily owing to the extremely uneven character of capitalist development, which provides a new economic sub-soil for the growth of exclusiveness, manifesting itself ideologically at the regional level in religious fundamentalism or in social reaction against minorities. The delicate balance between the legitimate aspirations of any individual nationality and the interests of the nation as a whole can be disrupted all too easily under conditions of extremely uneven growth. The fact that both unusual backwardness and unusual rapidity of growth can be conducive to a disruption of the balance is illustrated by Assam and Punjab, which have seen the rise of secessionism in different forms.

In Assam, a region rich in mineral resources but economically backward, the perception of being economically and culturally overwhelmed by Bengali-speaking immigrants gave rise to an exclusivist movement among ethnic Assamese which demanded the expulsion of those it labelled "foreigners", and also denounced "colonialism" by the central government on the grounds that Assam's oil serves the nation. There is hardly any indigenous Assamese manufacturing bourgeoisie; the activities of the industrializing state of India are perceived as alien by the intellectuals articulating a predominantly agrarian, petty-producer consciousness which gains expression as a right-wing radicalism: "radical" in its opposition to big capital and centralized authority, and right-wing in its scapegoating of identifiable minorities and its opposition to workers' movements. A family resemblance may be discerned with the agrarian radicalism of inter-war Japan.

In Punjab the origins of the secessionist ideology, and the subsequent terrorist movement against the central government and all organizations and individuals opposed to secession, lie in the opposite (but in its principles the same) phenomenon of extremely rapid agricultural growth which has made Punjab the most prosperous region in India, even though the bases for industrial growth lie elsewhere. The beginnings of Punjab's agrarian prosperity go back to the turn of the century, when most of the irrigation investment by the colonial state was concentrated in this region (part of which today lies in Pakistan). During the half-century before independence, Punjab which started far below Bengal in land productivity, rapidly overtook it and emerged as a wheat-exporting region. In the last four decades, as we have seen, Punjab has come to be identified with the Green Revolution and has emerged with a disproportionately high contribution to the country's food surpluses. The rural rich form a new restless stratum, "unemployed" because they are by and large unwilling to work on the land; at the other pole, nearly two-fifths of the rural workforce comprise agricultural labour, supplemented seasonally by migrant labour from the poor and backward state of Bihar some 1,000 miles away.

The burgeoning prosperity of Punjab's capitalist landlords and rich peasantry is buttressed by the considerable flow of remittances from Punjabis settled abroad (in Canada, the US and Britain). Punjab, along with Gujarat, has been for many decades strongly linked to the metropolitan capitalist world, and the link is directly from its villages and district towns, through migration. Economists and agronomists trained in US Mid-western institutions staff the local agricultural universities. At the same time there is a disproportionately small manufacturing bourgeoisie: numerous small capitalists running textile units or transport concerns do not add up to a capitalist class comparable in any way to the monopoly bourgeoisie which dominates the industrial economy in India. In Punjab the alliance between the landlords and manufacturing capitalists has been dominated on the whole by the landed interest with the ideology of capitalist modernization heavily biased towards a species of agrarianism serving primarily the interests of the increasingly commercially minded landlord capitalists. (These interests have ensured, for example, that imputed rent on owned land should be counted as part of production costs for the purpose of fixing the price at which government procures foodgrains.)

The landlord-capitalists' growing economic power has in turn generated, within a section of it, the aspiration for a greater measure of political power than it perceives to be possible within the "constraints" of the Indian union. The most extreme version of the demand for political power articulates itself in secessionism, i.e. the creation of a separate state on the basis of religion; the more moderate version which stops short of secession nevertheless asks for a degree of autonomy (all matters except defence and communications to be decided by the new state) which would effectively cut Punjab off from the Indian union as an economic entity. The terrorist movement is eagerly aided by a neighbouring country in receipt of heavy US arms aid. There is little doubt that if secession occurred, the first action of the proposed state of the landlord-capitalists would be to export food surpluses for profit, rather than supply the public distribution system within India; the second action would be to permit imperialist military bases to be set up. At the same time, the decades of the freedom struggle and the growth of democratic movements have generated a new consciousness among Punjab's masses, who have remained quite remarkably impervious to religious demagogy.

The "new agricultural strategy" was designed to overcome food supply problems by relying on private investment within an unreformed agrarian structure. In less than three decades, the very success of the new strategy in promoting capitalist investment has generated unprecedented problems of class and regional imbalances.

Notes

1. Schultz, T., *Transforming Traditional Agriculture* (Yale University Press, 1965).
2. Kahlon, A.S., and D.S. Tyagi, *Agricultural Price Policy in India* (New Delhi: Allied Publishers, 1983).
3. Blyn, G., *Agricultural Trends in India, 1891–1947: Output, Availability and Productivity* (University of Philadelphia Press, Pennsylvania, 1966).
4. In the nineteenth century, Ernst Engel discovered that as family incomes rose, the proportion of their budget spent on food declined. This effect has come to be known as Engel's Law.
5. Patnaik, U., *Peasant Class Differentiation: a Study in Method with Reference to Harayana* (Delhi: Oxford University Press, 1987).

8 Another Awkward Class: Merchants and Agrarian Change in India

Barbara Harriss

Our understanding of the effects upon society of the new agricultural technology in India has concentrated upon the sphere of production to the neglect of the sphere of circulation. Yet markets are the midwives of agricultural change. New technology in agriculture can be and has been adopted under conditions where land and labour markets are only partially and imperfectly developed: such markets catalyse or constrain rural and non-rural class formation. So too can markets for commodities: new inputs to the production process and increased supplies of agricultural outputs. Both food and agro-industrial raw materials are necessary for the development of the non-agricultural economy whose products in turn respond to the distribution of demand of the entire home market.

It is with markets for these crops that this chapter is particularly concerned. We investigate the relations of commercial capital and trade in three contrasting regions of India. The chapter shows how mercantile capital, buttressed by political dealing, siphons investible resources from agriculture, and may direct and control agricultural technical change and rural class formation. The diversity of these relations of exchange, and the common fact of the economic power of merchants, are central to an understanding of the diverse outcomes of the Green Revolution in the regions of India.

Three views about markets

Markets, being little studied, are vulnerable to having assumptions made about them. Extreme assumptions are attractive theoretically and politically. Markets can be assumed to be competitive and allocatively efficient, with traders considered at worst as a residual class and at best (not necessarily inconsistently) as entrepreneurs. This provides some ideological buttressing for a minimalist regulating role for the state. The reverse is also assumed: markets are inefficient and monopolistic, with the rural economy dominated by a class representing merchants' capital. Merchants' capital is deployed to appropriate surplus from direct producers either by itself or in fusion with modes of surplus extraction such as rent and usury, thereby tending to preserve old forms of production, relations of exchange and surplus appropriation. This gives justification for political intervention. State trading may be proposed to compete with and/or destroy private mercantile monopolies in the joint economic interests of allocative efficiency and of welfare or equity, and in the political interests of transforming relations and forces of production. Yet the third and common assumption about commodity markets is that they can be ignored. For in a strict sense merchants' capital is not a relation of capitalism. It is money used for buying and selling goods. Merchants, by specializing in buying and selling, are useful to society only by virtue of saving labour which would otherwise be dispersed in this necessary but essentially unproductive activity. By implication they are irrelevant to the process of class formation that has fascinated commentators on the Green Revolution.

All three views present caricatures of markets in the three regions of India where I have examined their operation. Despite a diversity virtually unique to the region in each case, whether imperfectly competitive or almost monopolistic, commercial capital is deeply entrenched in the rural economy. It is capable of influencing the ease with which new technology is introduced. It is involved in class formation and in structuring rural and urban capitalist development. It attempts to protect itself against threats to its independence so as to preserve the relations of commerce. In this process, it struggles over surplus with direct producers, with labour and with the state.

Paradoxically it is because merchants' capital is hardly ever to be observed by itself in a pure form that it performs the mediating roles described above. There are at least three sources of impurity of form. The first source is the combining of money used for buying and selling with productive activity necessary to the circulation of commodities: transport, processing (and also storage to the extent that resources are deployed to prevent goods from deteriorating). The second source of impurity derives from the ease of movement of merchants' capital into and out of productive industrial or agricultural capitalist enterprise according to relative rates of profit and to risk. In the regions examined, a straightforward historical process of subordination of merchants' to industrial capital is not evident. In fact complex commercial, financial, industrial and agricultural portfolios within which capital is constantly shifting are only too evident. The third source of impurity is the linkage of merchants' capital with usurers' or finance capital. It is argued by some that small-scale agricultural production rests on merchants' and moneylending capital which controls the reproduction of producers via money advances and crop pledges. Under such circumstances the lender can control what is produced, how it is produced and the extent and means of surplus appropriation, though the borrower retains operational independence, for mercantile control is never complete. The cultivator employing wage labour will be in a position to appropriate surplus value from that labour while simultaneously being embroiled in a process of surplus appropriation via interest on loans and via the price system. The severity of this latter, secondary surplus appropriation depends on the bargaining power of the producer with the moneylender or trader.

It is useful to distinguish abstract pure merchants' capital from concrete and impure commercial capital. While the former is independent of the mode of production, the latter may not be either independent of or determined by the mode of organization of production but may actually help to determine it.

This role can be illustrated by looking at the structure and functioning of agricultural markets, the appropriation of surplus by commercial firms, their control over the process of production and their role in class formation. Our evidence relates to the grain which dominates the economy of North Arcot District of Tamil Nadu in the south and Birbhum District of West Bengal in the north-east, and of grain, oilseeds, tobacco and cotton in the diversified dryland agricultural base of Coimbatore District of Tamil Nadu.

Structures of agricultural trade

Everywhere the structure of agricultural markets is complicated. The density of commercial intermediaries varies, being apparently crowded in surplus regions (an estimated 2,500 grain merchants' firms in half of North Arcot; an estimated 1,500 in a quarter of Birbhum) and less crowded in deficit regions (some 100 grain firms in a quarter of Coimbatore District, though there are 1,500 registered merchants in all, half of whom deal in cotton, the rest trading in fifteen other crops).

These relatively large numbers of agricultural commercial firms mask specialization and concentration in commodity flows and territorial control, and concentration and polarization in their size distributions. The richest 10 per cent of rice wholesalers in North Arcot trade forty times as much as the poorest 10 per cent. The top 1 per cent of traders in

Coimbatore control 20 per cent of trade and 9 per cent of assets while the lowest half control only 5 per cent of trade and 10 per cent of assets. The concentration of control over storage is even more extreme, the top 1 per cent controlling almost half the region's agricultural stores. In Birbhum District in West Bengal, a relatively pauperized rice economy supports a baroque and highly stratified edifice of commercial capital, labour and technology. At its apex is a handful of rice-milling magnates and at its base a crowd of (illegal) petty traders. The average rice mill does business 2,500 times larger than the petty trader and has assets 1,500 times larger. Oligopolies coexist with petty trade, the relationships between which change with the seasons, trade being less strongly dominated by the oligopolistic sub-sector after harvests and more dominated before harvests. The "restrictive practices" of the oligopoly perpetuate the wide trade margins that invite the small commercial firms; but the latters' capacity to usher in competition is constrained by their dependence on these very firms for credit for goods to trade, stores and transport to rent, information and contacts to solicit. The costly terms and conditions of the dependence of small firms upon large ones prevent the majority from growing big enough to be threatening.

Commercial firms are misleadingly classified in overlapping terms by volume ("wholesalers"), by position in the marketing chain ("retailers"), by process or function ("millers"). In fact these classifications are not mutually exclusive or watertight and many firms embody all three and occupy particular niches in the sphere of circulation. Nowhere is this more evident than in Coimbatore District. Here 95 per cent of firms engage in the distinctively mercantile acts of buying and selling, the remainder being pure brokers. Most firms also perform productive activity necessary to circulation: 85 per cent transport and 70 per cent process. But if their activities are spelt out in consistent detail, what emerges is a tremendous diversity and specificity of combinations of activities together with tendencies to functional complexity: 150 firms have 107 different combinations of activities. Most are unique. The agricultural markets are essentially heterogeneous in structure.

Draining surplus from agriculture

Modes of surplus extraction and redistribution are varied. Surplus is not only appropriated through buying and selling and through the internal production of surplus value, it is still accumulated in primitive, technically criminal ways: through manipulation of weights and measures, arbitrary deductions, closed transactions, delayed payments and adulteration practised on producers and consumers, through illegally low wages and casualization of wage labour in commercial firms, and from fraud and evasion against the state.

Agricultural markets appear to function so as to siphon resources away from agricultural production. Commercial capital accumulations are important in building and shaping the local non-agricultural economy. In Birbhum District in the north-west, agricultural rents and profits are important in the starting capital of small, recent firms. Agricultural rents are less important and profits from agricultural trading are more important in the much larger capitals of the established oligopoly. Profits from trading have been invested in commercial diversification (because of the striking seasonality of rice production and because of a trading environment made increasingly risky by government interference). Big trading firms have massive investment portfolios integrated not only horizontally into commerce in other agricultural products but also vertically into related agricultural industries. Direct investment in agricultural production is very rare. In Coimbatore District in the south, agricultural commerce provided over half the starting capital of agricultural firms and agro-industry, and trading yielded a further 20 per cent. Profit from agriculture or land sale was unimportant both as starting capital and as an investment for trading profits. The latter have been ploughed back into trade, used to splinter off specialized financial corporations or sunk into transport or property. It does not make sense to invest directly in land, its improvement and new agricultural production techniques, when rates of return to moneylending and commerce far exceed those to agriculture, and when the production of both foodgrains and agro-industrial crops is facing a squeeze in the relationship between (rising real) costs and (falling real) prices. In this region a secular decline in the profitability of much of agriculture can be traced. So much for the Green Revolutionary claim of lower costs per unit!

In North Arcot similar patterns and relationships can be discerned. Just under half of all starting capital had been accumulated from agricultural commerce itself, a fifth was borrowed, while a third was accumulated from agricultural production. Commercial profits were returned to commerce and moneylending or concretized in urban property, shops and stores. Commercial firms are gradually unhitching themselves from their agricultural origins and are lodging themselves in the urban economy. An ordinary grain-marketing town called Arni shows what is happening. While rice production doubled between 1973 and 1983, the flows of trade through the town increased by a factor of eight (discounting for inflation). Goods enter the town from increasingly long distances. The town is progressively more important as a wholesale centre for its rural region, and within the town economic power is being concentrated. The richest 10 per cent of firms have seventy times more assets than the bottom 50 per cent of firms. The richest 10 per cent own large tracts of the fabric of the town: houses, shops, workshops, sweatshops, fleets of lorries, finance companies, commercial firms and some factories. Rural inequality in income and purchasing power which has been the focus of so much critical analysis is dwarfed by rural-urban differentials. The income per head in an urban silk-manufacturing and trading household is 530 times greater than that in a rural labouring household. Yet that of the small minority of rich peasants and small capitalists who operate more than 2.5 acres (1 ha) is only ten times greater on average than that of the labouring household. The consumption patterns of the rich urban commercial households mould the development of national capitalism. The latter is highly spatially concentrated, fuelled by financial surpluses which flow via bank deposits and loans, via increasingly extractive flows of interest from direct producers to financial institutions such as private corporations and pawnbrokers, and via the investment of commercial cum financial profits. Financial surplus appears to flow from less developed rural regions to more developed destinations, with commercial activity in the metropolitan centres exerting a disproportionate magnetic pull.

The net drain of surplus from agriculture mediated by the market has not prevented revolutionary transformations in techniques of production, but they are uneven. In North Arcot a quite highly differentiated peasantry has been supplied by commercial firms with irrigation pumpsets, fertilizer and pesticides, and crucially with money and credit. In Coimbatore private traders acted in the same way much earlier as midwives to the cotton crop, but have had no material interest in the technical transformation of production of risky rainfed crops such as millet and groundnut. In Birbhum, while the agricultural commercial sector itself is technically and industrially innovative, conditions of grain production are technically relatively backward.

The unevenness with which commerce transforms production is not simply a matter of demand, it is also the result of the indirect control of the process of production by the commercial sector. The most powerful means of exerting control is through finance and usury. In North Arcot almost all commercial firms lend money for production and consumption to all classes of direct and indirect producers. Repayment is in grain at low harvest prices (lowered further to disguise interest payments). These financial relationships assure the merchants their supplies for speculation. The mass of the peasantry is compelled to borrow in order to increase production to repay an escalating burden of debt. Merchants have no interest in foreclosure because of the costliness of assumption of legal possession, the costs of direct management and supervision, and the relatively low rates of return from agriculture.

Likewise in Coimbatore, commercial control over cotton production is achieved through moneylending. In this case, however, the hierarchies of credit are more highly layered with subordinate agents (the largest 10 per cent of trader-moneylenders control two-thirds of all money lent); the web of credit is spun over a wider geographical region; the dealings are more confined to the kulak class of cotton growers and the possibility admitted of delinked cash repayments with interest for cash borrowed. Nevertheless, only a third of traders lend money for production and in this district there is great variation in the extent to which commercial capital indirectly controls production through moneylending. Traders dealing in rainfed foodcrops do not lend at all. There is little need for cash in the process of production. In the case of tobacco, very short-term preharvest loans are devices to secure perishable post-harvest supplies. The kulak class of tobacco producers actually reverses the process of financial control, and finances the long curing and highly volatile trading of their product. This class of producers also does contractual battle with client merchants over the rich but long-deferred pickings.

In Birbhum District, by contrast, a vast mass of tenants and pauperized petty producers are dependent for the reproduction of their households upon consumption loans taken under much more extortionate and individualized terms and conditions from the agents of a numerically small class of merchant-usurers and (former) landlords. Repayments in kind after harvest deprive direct producers of the means of subsistence and perpetuate a dependence which in turn preserves a structure of surplus appropriation and commerce.

The role of commercial capital in class formation reflects these regionally varied relations of control and surplus appropriation. In North Arcot, indirect commercial control, through moneylending, over the process of production slows that part of the process of depeasantization that would result from foreclosure upon default, assumption of possession and direct operation. Thus secondary surplus appropriation is constraining the development of large-scale capitalist agriculture based on primary surplus appropriation, in a region where both productivity and rates of return increase with scale. In Coimbatore, a frontier region with a long-established kulak class, commercial capital exerts a less thorough indirect control over production but gains very high rates of return through a combination of redistribution of surplus via buying and selling on the one hand and primitive accumulation on the other. Here, it is the rural proletariat who currently depend on commercial credit for the means of their subsistence. The ability of the commercial sector to exert definitive control over grain stocks and grain prices is as important in setting the terms and conditions of consumption of this proletariat as are the forces determining rural industrial and agricultural employment and wages. Finally, in Birbhum, the production relations termed semi-feudal by some (wherein surplus is extracted by rent and usury which combine to deprive direct producers of voluntary actions about production or exchange) have been transformed via tenurial reforms creating a petty commodity-producing peasantry. But the reforms fell short of non-land property, so that an unreformed and landed commercial sector sits tight, able to continue to extract surplus through a variety of modes: rent, usury, buying and selling, surplus value and primitive accumulation, which are the means of perpetuation of a backward, pauperized peasantry. They may now also be threatened by an influx of illegal small-scale firms attracted by very wide trading margins and appropriate to a post-reform, petty producing economy. This is the clearest example we have of relationships of mutual determination between production and commerce.

An awkward class

Merchants do not form a class of their own, not only on account of the intermeshing of productive and unproductive activity in the portfolios of those engaged in circulation, but also because of the heterogeneity of the mercantile sector wherein the redistribution of surplus from buying and selling to petty firms may be constrained by exploitative terms and conditions of financial relations binding them to large firms.

Their awkwardness resides in the fact that despite this lack of empirical distinctness, the oligopolistic sub-sector acts independently and politically in protection of its own interests, as a class for itself. Such action is expressed in ways more subtle than party politics, for no Indian party has a position on agricultural trade that is both consistent and promotional. Agricultural merchants are Vicars of Bray with regard to the government of the day. Their opportunistic funding of many political parties, not excluding communist parties, is a risk-minimizing tactic. The prominence of agricultural merchants in local institutions of co-operation, government, religion and philanthropy consolidates power alliances. Their political institutions (commodity associations) have sprung into existence in response to their own needs for autoregulation, and to threats from labour and from the state. Where industrial capital is thickly spread (as is the case with the cotton trade and the textile industry) these lobbies have been found to align themselves speculatively according to perceptions of the distribution of benefits at a given conjuncture. Where productive agro-industrial capital is thin, these lobbies tend to react so as to preserve the independence of the commercial sector. They exert certain influence over local government, judiciary and police.

The conflict between the regulating state and regulated markets is more apparent than real. There are strong mutual interests between merchants and the state. Both benefit from the misimplementation of regulatory interventions. Merchants benefit through (illegal) profits derived from distorted or black markets. Low-level bureaucrats benefit from "institutional rents": bribery and corruption. There is, furthermore, much compromising recruitment of merchants into state trading institutions, directly as advisers or indirectly as agents. The commercial sector succeeds in receiving greater subsidies than is apparent from planning rhetoric and other statements of intention by the state. All such relations are made the easier by increasingly close kinship ties between merchants and the bureaucracy.

The state acts ambivalently towards the commercial sector. In North Arcot it both regulates and competes with the private grain trade – most imperfectly, in both cases. State support to peasant production has accompanied the widespread transformation of production techniques and a doubling of production within a decade. In West Bengal, peasant forms of production have been strengthened through state-initiated land reform, while rice production languishes if compared with North Arcot's. It is not clear that such relative technical backwardness is the result of the creation of a petty commodity-producing agrarian society through land reform, or whether it results from lack of reform of the highly concentrated state-protected commercial sector upon which the state relies, or whether instead it results from intractable environmental factors and could in principle be "technically fixed".

In all our cases, including the capitalist agricultural and industrial district of Coimbatore, the agents of the state profit privately from the indirect control over market-orientated production perpetuated by commercial interests. This nexus of interests siphons resources away from direct producers. Resistance to such economic and political relations can hardly be sought in state institutions like co-operatives and other purveyors of credit and regulatory and fiscal measures which are dominated, captured or evaded by merchants. Instead it awaits challenges from capital and labour. Challenges from capital are more likely to be issued from farmers' organizations (the nearest thing to class for itself existing in a polity shattered vertically) than from the aptly named (urban, "industrial") Chambers of Commerce. Challenges from labour are more likely to come first from trade-unionized labour forces within mercantile firms than from agricultural labour unions. At present such challenges seem to be only weakly developed.

Background Reading

Bhaduri, A., *The Economics of Backward Agriculture* (London: Academic Press, 1983).

——"Forced commerce and agrarian growth", *World Development*, vol. 14, no.2 (1986) pp. 245–55.

Harriss, B., *Transitional Trade and Rural Development* (New Delhi: Vikas, 1981).

——*State and Market* (New Delhi: Concept, 1984).

——*Agrarian Change and the Mercantile State* (Madras: Cre-A, 1985).

——"Merchants and markets of grain in South Asia", in *Peasants and Peasant Society*, ed. T. Shanin (London: Blackwell, 1987).

9 Agrarian Crisis and Political Crisis in Mexico

Roger Bartra

The Agrarian Leviathan

In Mexico, the peasants have suffered from a widespread belief that they owe their existence to the state, and that they should therefore do whatever is called for by the president. Although the memory of President Lazaro Cardenas, who changed the life of millions of peasants at the end of the 1930s, has faded considerably, the agrarian measures introduced by his government still exert considerable influence on the rural political situation. With the benefit of hindsight, it is possible to see that the agrarian reform was necessary to enable capitalism to develop in the countryside. The *latifundias* were acting as a break on modern capitalist development and were responsible for uncontrollable rural misery.

The agrarian reform called for an egalitarian distribution of land and provided small plots to many peasants, but it carefully maintained a small group of commercial farmers who rapidly came to own a large part of the wealth produced in the countryside. In the same way that the bourgeoisie does to workers, the agrarian reform placed the peasants in an unequal struggle. It gave them a tiny piece of land but left them facing the voracious capitalist system. The peasants had lost the battle before it began.

The agrarian reform also had a very important political function: to incorporate the peasants into the political system, as a huge and passive popular base for the official governing party. The results were spectacular. Twenty years later proud government politicians were still presenting Mexican agriculture as an enviable model which had succeeded in combining accelerated economic development with political tranquillity.

However, the 1970s and 1980s saw a profound crisis of agriculture and a striking increase in protests by numerous different groups in the countryside. There were various forms of protest, from guerrilla struggle and violent land occupations to strikes, peasant marches to the cities, and stoppages at the agricultural companies. In the last twenty years the situation in the countryside has changed completely, and both political peace and the growth of production have come to an end.

The double crisis

Let us first briefly examine the economic characteristics of the crisis. It is a complicated situation but it can be understood as an interlocking spiral of two different crises. We find ourselves with a contradictory relation between the accelerating expansion of the capitalist sector and the specific features of the peasant economy. Since the 1950s, capitalist development has been seriously eroding the small-scale economy of the peasants, which was one of the bases of stability in the countryside. Between 1945 and 1956, agricultural production had grown at 7 per cent a year, but by the end of the 1950s, the rate of growth had fallen to an average of 2.5 per cent a year.

Poor peasants contributed a large part of the agricultural product, and the ruin of the peasants led to a far-reaching agricultural crisis. The capitalist sector did not produce enough to feed the population: it had orientated its production to commercial crops for export, inputs for non-food industries, and luxury or non-basic foodstuffs. Capitalist agriculture was incapable of supporting the system without relying on imported foodstuffs, and, as hundreds of thousands of peasants were ruined, the national production of food was reduced.

It was with the aim of checking these tendencies that important distributions of land occurred in the 1960s. The process was paradoxical, since it was governments that were not exactly populist, like that of Diaz Ordaz, which revived the old agrarian policy. However, this did not stem the erosion of the small-scale peasant economy: on the contrary, it became more acute. This happened in part because the land that was distributed to the peasants was of very poor quality. Although the government of Diaz Ordaz (the same one that massacred hundreds of students in 1968) distributed 25 million ha of land, only 10 per cent was arable. By contrast, in the golden era of land distribution, President Cardenas distributed 20 million ha, but 25 per cent was arable.

In addition to the crisis caused by the ruin of the peasantry, in the 1970s there was also a fall in the output of the modern capitalist sector of agriculture. This was characterized by a decline in prices which was preceded by a typical period of overproduction. During the decade there was a marked decline in the production of sesame, cotton, sugar cane and tomatoes. In some cases, as with cotton, the crisis appears to be long term, and is the result of developments in the world market. But in general terms, the crisis of the modern sector has a cyclical character and depends to a large extent on the fluctuation of prices.

The agricultural crisis has its roots in the 1950s and happened principally because capitalist development in agriculture entered a monopolistic phase with an inadequate economic and political environment. The concentration of capital and of land, the penetration of financial flows and the monopolization by the state of sectors of production and marketing occurred within a juridical, social, political and economic framework that restricted modern capitalist forms of exploitation. The "reformist" phase of capitalist development in the countryside was checked long ago; its final, weak signs were the distributions of land in the 1960s, which instead of developing production helped to precipitate the crisis.

Populism and efficiency

During the 1970s, and in response to the double crisis that was occurring in the countryside, the government implemented a twofold policy. By means of a Law of Agricultural Development it attempted to open the reformed sector to the investment of capital. The reformed sector consists of land that is not part of the regime of private ownership but rather of the system of *ejidos,* which is a mixed system of nationalized and communal property (based on the principle that "the land belongs to those that work it"). This law openly permitted the letting of plots of *ejido* land, as well as the employment of waged labour on land that had been distributed "to those that worked it". Now the land will not belong to those that work it, but to those who cause it to produce: a shift that marks the difference between the battered peasant economy and the capitalist economy. Thus, the government of Lopez Portillo implemented changes in the 1970s which I would rate as a thoroughgoing dispropriation of the peasants' property. As with the communal indigenous lands that were affected by the Lerdo law of 1856 (or the famous Enclosure Acts in Britain), the government of Lopez Portillo pursued a policy intended to encourage capitalism to take over the production which the peasants were not capable of developing.

However, Lopez Portillo's government did not completely abandon the traditional agrarian rhetoric. With the aim of "revitalizing the alliance between the peasants and the state" and of solving the country's food problem, it established the Sistema Alimentario Mexicano (SAM). Under the guise of a food Utopia, SAM never amounted to more than a policy for prices and the distribution of inputs which was designed to raise the production of basic grains and to strengthen the strata of peasants that had been incorporated (rich farmers, of course, also benefited from this policy). The creators of SAM recognized that some producers would fall outside their development plans and "would not be able to survive as farmers". This part of the rural population – "below subsistence" peasants with less than 2 ha – amounted to about one million producers in 1970 (not counting their families). This is more than one-third of the total. What did SAM propose for them? It suggested combining their plots of land and offering them employment and subsidized consumption in other spheres of the economy.

This twofold policy, which interwove a faded populism with technocratic efficiency, has been followed in the 1980s by the government effectively abandoning the countryside to the tendencies outlined above. It can be said that the government has no agrarian policy, and is concerned principally with surviving the most serious economic crisis the country has experienced for fifty years. The reformist and populist vein is exhausted, but no alternative has been developed. How is it possible to be efficient in the middle of a crisis that corrodes the whole system?

The political crisis

Perhaps one of the most interesting aspects of the current agrarian problem is that it has helped to trigger an acute political crisis. The Mexican political system has depended to a large extent on the rural population. In fact, the system was developed when Mexico was a fundamentally agricultural country. During the 1930s and 1940s the peasants were positioned by capitalist logic in an intermediate position between the two fundamental, antagonistic classes, the proletariat and the bourgeoisie. Their role was supposed to be both political and economic: to be a balancing factor in class conflicts and somehow to contain the labour force that the economy could not employ in industry or services. But today things have changed: in the context of a predominantly urban nation, the peasantry is being proletarianized and monopoly capital (both private and state) has a decisive presence in agriculture. These new conditions, combined with the agricultural crisis, have created a need to reorganize the reformed (*ejidal*) sector of agriculture. The possibilities for reorganization range between two extremes: (a) to permit the free circulation and concentration of capital in the *ejidal* sector, and (b) to direct the concentration of capital in a form that is controlled and financed by the state as *ejido* collectives, co-operatives and/or decentralized state enterprises.

In recent years there have been a number of skirmishes between different fractions within the dominant class over which alternative should prevail. Although it has not been acknowledged explicitly, it appears that tendencies closer to the second alternative are tending to predominate. However, whatever form the concentration and centralization of capital might take, there is no doubt that the future of the peasant economy will play a key role in determining the economic and political balance in the countryside.

The dominant classes are faced with the problem of dissolving the peasantry without provoking political chaos while at the same time finding an alternative capitalist path of development. The erosion of the peasantry implies the destruction of a crucial part of the political system's social base: the network of *caciques* (chieftains) and the National Confederation of Peasants (CNC), which is an integral part of the official party.

The *caciques* are losing power for two main reasons. First, their power is based on mediating between the community and the government; if the community is destroyed, they will cease to have any purpose. Second, many of the new modernizing, technocratic members of the bureaucracy are repelled by the *caciques* and look upon them as relics from the past. For similar reasons, the CNC is losing power. As the reformist tendencies of the state have been exhausted, the CNC's function has ceased to be significant, and the control which it used to exercise over the rural masses has been weakened. In addition, the conservative currents in the government were hostile to the populism of the peasants' leaders, and were keen to reduce the importance of their organization.

The dispossession of the peasantry has affected the whole of society. It has provoked a runaway growth of the so-called marginal sectors, generating violence against private landed property and causing an intermittent flow of migration to the United States and the large cities. The political system was not prepared for the avalanche of conflicts and problems that arose as a result of the extension of capitalism in agriculture.

This does not mean that the fundamental cause of the political crisis lies in the collapse of the peasant economy and the consequent process of proletarianization. But I do want to emphasize the great influence that these factors have held in changing the rules of the political game. The socio-political dynamic has shifted from the country to the city; however, the rulers have still not accustomed themselves to this and accepted that it is no longer so easy, for example, to call on the rural vote to legitimize the system in the face of the growing opposition of important sectors of workers or of the middle classes in the cities.

Within the official party a current has arisen that calls for changes in the system of government. This current is headed by political leaders linked to the populist traditions of the state, and originally included Cuauhtemoc Cardenas, the son of the great agrarian president. Cardenas has since broken with the official party, and stood for president against its candidate in the 1988 election, mounting the first serious electoral challenge to the official party since it was formed in the aftermath of the Mexican revolution.

The political system is rather like a three-legged table. It was underpinned by three classes: the bourgeoisie, officially called the middle class: the workers: and the peasants. The institutional continuity of the state rested to a large extent on the peasant base of the regime. The peasants, the class linked most symbolically to the Mexican revolution, have become a weak and unsteady base. The ruin of the peasantry is not just the end of the revolutionary myth that bolsters the system; it also gives notice of the ruin of the system itself.

10 "Structural Adjustment", Agribusiness and Rural Women in Tanzania

Marjorie Mbilinyi

A prominent World Bank official stated recently that the Bank's new policy aim should be to shift agricultural production from the high-cost zones of North America and Western Europe to low-cost zones like Africa.[1] This is consistent with American efforts to reduce the support that maintains family farming in Europe,[2] and to bring the continent into line with the USA, where the number of working farm families has halved in the 1980s. The Bank's view implies "runaway" plantations and agribusiness schemes (see Watts's chapter in this volume), like the runaway factories that have relocated from America and Europe to the Third World in search of cheap labour and other attractive conditions.

In the case of Tanzania, the historical and contemporary significance of large-scale farming and agribusiness – and its effects for peasant farmers and rural workers – has largely been neglected. Starting with a critical commentary on the currently dominant World Bank/IMF view of Tanzania's economic, and especially agricultural, crisis, this chapter then demonstrates the major role of large capital, especially TNCs (transnational corporations), in Tanzanian agriculture and the scale of casual labour. It then argues that in rural Tanzania, the views and struggles of rural women are central to the outcome of the World Bank's efforts to engineer its chosen "solution" to agrarian crisis.

The mainstream view of Tanzania's agricultural crisis

In the mainstream view, shared and broadcast by the World Bank, the crisis is defined by declining crop production and export earnings from agriculture during the 1970s and 1980s. The data used to support this view are very suspect "guesstatistics", as we say in Tanzania, largely derived from crop sales in official (state-controlled) markets.

Yet a substantial proportion of export crops and food crops are exchanged, and circulate through other channels including smuggling across borders (coffee, maize, beans, potatoes, cardamom) and sales of food staples in local and "parallel" markets. A decline in the volume of officially recorded sales does not necessarily reflect trends in production or exchange.

Second, the data used are often highly aggregated and fail to distinguish sufficiently between crops, or to distinguish at all between peasant and large-scale farming. These aggregated "guesstatistics" obscure the extreme unevenness of trends *within* peasant farming, and *within* large farming. There have been major declines in peasant-grown cotton and cashew nuts, with an increase (even in official sales) of coffee, tobacco, smallholder tea, maize and other grains. Estate production of sisal, once Tanzania's major export earner, has been declining since the mid 1960s, while more recently sugar production and sales have dropped massively despite major funding by the World Bank and other aid agencies to the (public-sector) sugar estates. On the other hand, and despite fluctuations, sales of green tea, paddy (rice) and wheat from large farms increased during the 1970s.

A third point is that per capita calculations/estimates of agricultural production based on total population, rather than population engaged in farming, can be seriously misleading. We do not know enough about this, but recent local studies suggest that fewer people now engage in farming, and *an increasing proportion of farm labour is female* (although women resist the terms on which they are incorporated in agricultural work, as is shown below). This questions the usual simple assumption of declining per capita production in agriculture, despite the lack of adequate investment in peasant farming, and suggests the intensification of (especially female) labour.

Mainstream explanations of "crisis" range from the "backwardness" of peasant agriculture (and peasants) to state policies, above all crop pricing and marketing policies which "tax" peasants and inhibit incentives (emphasized by the World Bank). "Interference" by the state with the "natural" working of market forces includes: investment in "non-productive" social services like schools, rural water supplies and rural medical clinics; import–export regulations to protect domestic industry and trade, including agriculture; currency regulations; price controls and rationing of scarce commodities; "uneconomic" regional policies including distribution of farm inputs and equipment to poorer areas; "uneconomic" labour policies.

This list shows that any redistributive policy, or policy aiming to meet basic needs, is likely to be condemned as economically "irrational". While state pricing and marketing policies have undoubtedly had some negative consequences for some or most peasant farmers, the mainstream view ignores resources returned to rural areas in the form of services which peasants, like workers, view as their rightful due, and as a recognition of their labour. Such services are especially important for those many rural women with little or no cash income (including those who help to grow cash crops, the proceeds of which are appropriated by their husbands or other "senior" men).

Again, with all their imperfections, price controls and rationing were introduced to combat shortages of basic commodities, and high prices and profiteering on the black market. The late 1970s and early 1980s saw a rapid expansion of women's co-operative shops, for example, which received priority in the allocation of basic goods.

The most glaring "silence" in the mainstream view concerns how the World Bank and other aid agencies themselves contributed to Tanzania's agricultural crisis, which the Bank conveniently blames on state policies. As Cheryl Payer has noted, "From 1974 onward the World Bank became deeply involved in nearly every aspect of Tanzania's agricultural system", including its consistent advocacy of export agriculture.[3] While the Bank claims that state policies discriminated against agriculture, large-scale farming and marketing agencies received a growing share of development finance in the late 1970s and 1980s, much of it from the World Bank.

Capital in Tanzanian agriculture

The significance of large-scale production in Tanzanian agriculture is usually ignored or denied. In fact, a sizeable proportion of the marketed value of agricultural commodities was produced by large farms in 1980 (see Table 10.1). Public or private estates produced 75 per cent of marketed green-leaf tea (private TNCs); 100 per cent of the sisal (half private); 100 per cent of wheat (public); 85 per cent of sugar (public); and 50 per cent of rice (public). Moreover, all the public corporations involved in large farming are dependent on TNCs for management and other services, for technology and other inputs. Some are being privatized now as part of the World Bank–IMF Structural Adjustment Programme (SAP) which began nationally in the early 1980s.

As well as owning their own estates and plantations, TNCs like Brooke Bond, George Williamson, Amboni Estates and the Karimjee Group monopolize upstream and downstream activities in peasant and state farming systems as well. For example, TNCs dominate the provision of feasibility and management services, farm inputs and equipment, turnkey factories, transport, packaging, insurance and marketing in Tanzania as all over Africa. Their activity expanded during

Table 10.1: Percentage of total crop marketed in Tanzania by type of production organization

Food crops	*Peasant (under 10 ha)*	*Medium (10–100 ha)*	*Large (100 ha +)*	*Private estates*	*Public estates*
Maize	85*	10*	5*		negligible
Rice	50				50
Wheat			——5[a]——		95
Drought staples	95*	5*			
Sugar	15[b]				85
Legumes	90*	5*	5*		
Export crops					
Coffee	85[c]			10	5[c]
Cotton	95*	5			negligible
Sisal				50	50
Cashews	100				
Tobacco	90	5*		5[d]	negligible
Tea	25[c]			70	5[c]
Pyrethrum	100				
Seed beans		——100——			

Notes:

* rough estimates; no precise breakdown available
[a] formerly (early 1970s) over 90% of official procurement
[b] peasant outgrowers at public estates
[c] breakdown between smallholders and public estates estimated
[d] formerly (early 1970s) 25% of the total

the 1970s as a result of "aid" from the World Bank and other agencies. A large proportion of this funding provided profits for TNCs by purchasing goods and services from them at inflated prices. Given declining opportunities in North America and Western Europe for agro-industry, bank loans to Tanzania and other African countries provided significant support to its activities.

There is also a small but growing sector of individually owned capitalist farms in Tanzania producing beans, horticultural products, maize, coffee, livestock and livestock products (dairying, pigs, chickens). This sector has been encouraged to expand by the "new agricultural policy" which accompanied the SAP, and a growing number of wealthy Tanzanians are investing in farms of 100 ha or more. However, they cannot compete with the TNCs in exports, due to their smaller scale, impaired access to foreign exchange to import farm inputs, equipment and technology, and to overseas markets dominated by the TNCs themselves.

Labour in Tanzanian agriculture

Agricultural production in Tanzania is based on various labour regimes: regular labour, migrant labour, casual labour and peasant labour. These developed initially during the colonial period and have persisted to the present, with modifications. During colonialism sisal estates were dependent on predominantly male migrant contract labour. Tea and coffee estates and mixed farms relied primarily on local squatter casual labour, much of it female, especially in tea and coffee harvesting. Peasant labour was engaged in both subsistence and commodity production. Some important cash crops, notably cotton, cashew nuts and half the coffee output, were largely produced by peasant farmers and absorbed much female family labour.

Sexual divisions of labour in rural households changed as a result of the migrant and casual labour systems, and the growing commoditization and proletarianization of the rural economy. One result was the intensification of female labour in both subsistence and market-orientated farming, a trend continuing after independence as a result of the expansion of the male labour market and increased cash needs forcing women to work harder to maintain themselves and their families.

Employers found it costly to hire permanent workers after independence because of the rights and benefits won by organized labour. Casual labour continued to provide an alternative for employers in agriculture. Casual workers among those employed in the organized sector increased during the 1960s and early 1970s, and then stabilized at about 25 per cent. The proportion of enumerated workers who were women also rose from 8 per cent in 1973 to 15 per cent in 1980. The policy of Universal Primary Education from the mid 1970s increased the numbers of female primary school leavers in the labour force, who are paid less than their male counterparts.

However, a barrier to cheapening the cost of labour power in agriculture is the expansion of "off-the-books" (unregistered) activities which provide many urban and rural women with their cash income. The crisis of declining crop incomes and wages which rural households have faced since the 1970s generated a shift of female labour from farming to other activities such as beer brewing, food processing, petty trade and prostitution. This move also expressed resistance by rural women and youth, both male and female, to oppressive patriarchal relations in households and villages. Increasingly, peasant women and young people have refused to provide unpaid labour for husbands, fathers and other household "heads". This has major implications for agriculture given the high female:male ratio of the rural population, and of the agricultural labour force (peasants and farm workers). According to the 1978 population census, some 63 per cent of this labour force aged 15 to 29 were women.

The expansion of non-farm commodity activities was further stimulated by the rapid expansion of the black market during the late 1970s and early 1980s, which also generated increasing social differentiation in rural areas. In the banana trade in Rungwe (Southern Tanzania), for example, a few women have become rich but most remain poor traders and worker-peasants, barely able to earn enough to maintain their families. By the early 1980s, many tea growers once categorized as kulaks (rich peasants) were impoverished and struggling to feed and clothe their families. As one man said, "Before we used to farm to get rich, now we farm merely to subsist."

The basic cost of living has risen dramatically in the 1980s with liberalization, devaluation and other aspects of World Bank/IMF conditional lending, and this has driven a growing number of people – especially the youth – into the casual labour market. However, they seek work in smallholder agriculture or in non-farm "off-the-books" activities whenever possible, rather than on estates and plantations.

Rural women speak

This section draws on statements made at "speak bitterness" sessions that Mary Kabelele and I organized in Uyole, Mbeya region, in late 1981 and early 1982.[4] The meetings were arranged by the Uyole farmers training programme during the preparation of a plan for regional integrated development. Women participants converted these meetings (intended to elicit their views about development goals and strategies) into a forum to voice their criticisms and complaints to regional, district and village leaders as well as fellow villagers.

Running through nearly all their statements was a protest against the sexual division of labour in its broadest sense: relations governing access to, and control over, land, labour, cash, farm inputs and equipment, the division of farm income, and organization of the labour process with its differential work loads for women and men. Women noted the intensification of female labour in farm and off-farm activities. "We work from morning till night" was constantly said, or "Transporting harvest from the farm on the head is difficult because the farms are far away." Women recognized their double burdens: "Women farm, they care for children, they purchase clothes for the children"; "Cultivating by hand with a baby on your back – that is a problem."

Conflicts over labour between individual household farms, village farms and separate women's co-operatives were noted. It was impossible for women to contribute to all of these activities, and also carry out domestic labour like carrying water, collecting firewood, and food processing and preparation.

Of great concern was women's lack of control over the distribution of income at household and village level. In Mbalizi Village, someone argued, "The husband is the boss! He decides how much cash to give the wife from the coffee." In Msia, a debate arose over control of village resources. One group of women insisted that they lacked access to the village tractor, and refused to be silenced by the local leader of UWT (the national women's organization) who contradicted them. In Isongole, women argued that they helped build the village school and the local party centre, but did not receive reciprocal help from male villagers or the village government to build a women's co-operative house.

The scarcity of basic goods like soap and clothing and the rising cost of living were other common complaints. In Halungu Village, people criticized government by pointing out the inadequacy of village medical, water and transport services. One elderly woman even presented an impromptu skit to dramatize her words:

> "If we have a sick child, we walk miles and miles on foot" (walks slowly, step by step, carrying an imaginary child in her arms, who gets steadily heavier as she goes along). "We go to a rich person to help drive the sick child to the hospital. He says that there is no petrol" (gestures as if imploring someone to help them; turns away in disappointment). "By the time we get to the dispensary, the child has died" (dramatizes the discovery that the child

> has died; shock, then mourning). "We return home" (has turned around as if to return in the direction she came from).

According to Tanzanian villagers, the provision of social services like health clinics, clean water and schools is *owed to them* by the government in return for the export crops which they produce. Their logic is this. They are constantly told by state functionaries that they must produce more export crops so that the nation may earn foreign exchange, and more food crops so as to reduce food imports and save foreign exchange. Foreign exchange, they are told, is needed to purchase medicine, agricultural equipment and other imported goods and services. Peasants cannot eat foreign exchange, nor do they directly receive a share of their own foreign exchange earnings (although MPs recently voiced village co-operatives' queries as to whether they also had a right to control 50 per cent of their foreign exchange earnings, the same as TNCs and other large-scale agricultural producers – another part of the Structural Adjustment Programme). Instead, peasants expect to share in benefits like social services paid for in part by the foreign exchange which they have "produced".

Women spoke out vigorously against patriarchal authority. The following speech by a female ward secretary of the party in Halungu in 1982 is an eloquent expression of the many issues raised.

> Women can no longer tolerate these oppressive conditions. It is so bad that they decide to leave. They leave their children, the husband, the farm, the house, to go follow the life of a prostitute. . . .Look at the women! They have lost weight this season because of work. The woman says, "I have the right to be self-reliant, but what will my children do?" She agrees to return home.
>
> Women work on the village farm, but very few men do. Women weed the coffee, they pick coffee, pound it and spread

> it to dry. They pack and weigh it. But when the crop gets a good price, the husband takes all the money. He gives each of his wives 200 shillings and climbs on a bus the next morning. . .most go to town and stay in a boarding house until they are broke. Then they return and attack their wives, saying, "Why haven't you weeded the coffee?" This is the big slavery. Work has no boundaries. It is endless.

Some women leave their husbands, children and household farm, and go to town. Others remain, but refuse to work without payment on cash crop production under their husbands' command. In Rungwe, men were forced to share the proceeds of the tea harvest after women and children refused to pluck tea on the family tea farms – they went to work for neighbours instead. And as noted above, women have tended to move into "off-the-books" activities outside the household domain, and therefore become independent of their husbands, fathers and other household heads' control.

What impact have women's resistances had on estate and smallholder agriculture? We know little about the interaction between estate and peasant farming. However, many women refuse to work on estates because of inadequate wages and oppressive work conditions, just as they refuse to work, or carry out "go-slow" actions, on family farms. This may help explain the ***shortage of casual labour*** which the major estates complain about.

According to tea, sugar, sisal and pyrethrum growers labour shortage is a principal cause of their production problems. Since historically women provided a substantial share of casual workers, their current resistance and the extent of their ability to survive outside farm work no doubt contribute to the alleged crisis of estate labour. However, what is at issue is the unwillingness of TNCs and state corporations in agriculture to pay adequate wages, and to provide improved working and living conditions for regular and casual workers.

The concluding section discusses attempts to restructure

labour, and in particular to contain the resistance and options of women, in present conditions of World Bank/IMF lending.

The "structural adjustment" of capital and labour?

In the crisis of the 1980s most African countries have been forced to accept Structural Adjustment Programmes; that is, desperately needed loans are conditional on implementing policy "reforms" dictated by the World Bank and IMF.

In the case of Tanzania the state has been told to deregulate import and export trade in favour of TNCs and other large producers. TNCs are to control their own foreign exchange earnings: to be able to siphon their profits out of the country without restriction. Investment in marketing and transport, in farm inputs and equipment, is to be concentrated in the most productive regions, districts and enterprises. The removal of fertilizer subsidies primarily harms peasant producers.

Devaluation of the Tanzanian shilling works in favour of TNCs by reducing local production costs. At the same time it increases the costs of imported farm inputs and equipment, and channels them towards large-scale producers. The removal of consumer subsidies for maize, and the reduction in price controls for basic consumer goods, contribute to the soaring costs of living for workers and many peasants who have to purchase part of their food needs. So do the introduction of "user fees" for education and health (part of the World Bank's programme), and the rising costs of these services.

What is at stake is an attempt at the structural adjustment of capital and labour in favour of capital. Key elements of this strategy include the promotion of export crop production as against self-sufficiency in food, dependence on the world market for imported manufactures as well as food (in effect, deindustrialization), and the promotion of agribusiness as against peasant farming. "Rural employment generation" is the answer to declining incomes from peasant farming, and means a more reliable pool of casual labour for large-scale agriculture.

As so often with strategies of capital to produce a cheap and compliant labour force, all this implies a partial and selective proletarianization, not least with respect to gender. The World Bank is aware of the significance of female labour in peasant farming, and its awareness is now shared by other aid donors and increasingly by African governments. The attention being devoted to rural women suggests that they have been targeted as a problem in need of a solution. Concern is expressed about the movement of women out of farming, their migration to towns, and increasing rates of divorce and single motherhood.

I believe that Women in Development (WID) programmes like income generation projects and training schemes which invariably target *rural* women, are intended to keep women in the countryside. Sometimes this is combined with legally enforced barriers to urban migration and settlement, attempted in Tanzania and several countries in Southern Africa.

Even in the short term, the effect of WID projects is to intensify female labour and the exploitation and oppression of women, since these projects are inserted in existing structures of power rather than challenging them. Such structures include "customary" laws of inheritance, marriage and family, which reproduce male domination in households and villages through patrilineality and patriarchy. Women remain dependent on men for access to land and other productive resources, housing and other necessities of life.

Lower wages and incomes make economic independence and survival more difficult for women as individuals, and especially single mothers, than for men. Typically women are forced to reside with a male partner, with parents, older siblings or other "guardians". This may help explain the significant rise of female "unpaid family workers" in both rural and urban areas during the 1970s.

The circumstances of the "structural adjustment" of capital and labour in the current African scenario of the World Bank and IMF include the immense pressures on the incomes of peasants and workers, and the possible decline in alternatives to wage labour resulting from liberalization and other SAP measures. For example, opportunities for petty trade in rural areas may decline as larger traders move in, as villagers in Rungwe observed in 1987. Nevertheless, the resistance of workers and peasants, not least the resistance of rural women to exploitation by both capital *and* patriarchy, may well "bog down" the smooth implementation of new policies for agriculture, as their advocates put it.[5]

Notes

1. Schuh, Edward, in a talk on "Agriculture and food policies" given in Brussels, April 1987.
2. See World Bank, *World Development Report 1986* (Washington, DC).
3. Payer, Cheryl, "Tanzania and the World Bank", *Third World Quarterly,* vol. 5 (1983).
4. See Marjorie Mbilinyi with Mary Kabelele, *Women in the Rural Development of Mbeya Region* (Dar es Salaam: Tanzania/FAO, 1982). The arguments and evidence underlying this article are developed more fully in Marjorie Mbilinyi, "Agribusiness and casual labour in Tanzania", *African Economic History*, vol. 16 (1986); "Agribusiness and women peasants in Tanzania", *Development and Change*, vol. 19 (1988); and *Big Slavery: The Crisis of Women's Employment and Incomes in Tanzania* (Dar es Salaam: Dar es Salaam University Press, forthcoming).
5. The phrase is taken from Mandivamba Rukuni and Carl K. Eicher, *The Food Security Equation in Southern Africa* (Michigan State University: Department of Agricultural Economics, Reprint 5, 1987). Carl Eicher is a senior USAID adviser on food security in the Southern African Development Co-ordination Conference (SADCC) region.

11 Hunger and Women's Survival in a Bangladesh Slum

Jane Pryer

Issues about food concern *access* – the ability to obtain food – as well as agricultural production. One of the most serious problems blocking access to adequate food is the increasing poverty of women, or the "feminization of poverty". Bangladesh, probably as much as anywhere in the world, demonstrates this process in all its severity.

It has one of the fastest rates of urban growth in the world at around 10.6 per cent a year. Urban migration has been mostly by men in search of work, and the major cities of Bangladesh have very high proportions of male to female inhabitants. Recently, however, there has been an increasing trend of independent migration by rural women, as a result of the feminization of poverty in the countryside.

Power and resources are distributed very unequally even within urban slums and their typically impoverished populations. This paper shows how poor women struggle to survive, drawing on research in the city of Khulna. It also suggests more generally that access to food is as much an urban as a rural question in Third World countries, with their marked inequalities of class and gender.

Khulna and Medja Para

Khulna is the second port and third largest city of Bangladesh, with about 750,000 people in 1986. Historically it has been an important regional trading centre, with a modest industrial sector. Over half its population live in slums and squatter settlements, with some of the worst unemployment and income levels of any city in the country.

Since 1950 the slums have spread from the railway tracks in the old part of the city to the commercial and industrial areas and beyond. Medja Para (a pseudonym for the slum described here) is near the main commercial area, and many of its residents are petty traders and labourers. Medja Para, then, is an established inner-city slum. It covers an area of 1.8 acres in which about 2,200 people are crowded together. Slum land and property on it is claimed by eighteen landlords, although disputes over boundaries and land titles are common.

All landlords live in the slum in *pucca* (cement-built) housing. Most of their households contain large extended families. The landlord families make their living by a variety of means in addition to rents – through trade, skilled employment, agriculture and moneylending. They are able to get low-interest loans from banks for investment, and their incomes are three to seven times above a local slum poverty line (SPL) constructed by the author.[1] The largest of these landlord families has a history of local political dominance over generations.

Struggles for survival of the poor

Contrasted with the relative affluence of the landlord families, the majority of slum dwellers are extremely poor. Most of them have been in Khulna for a long time (on average twenty-two years), having migrated from the surrounding countryside because of rural poverty.

Wage labour and petty trade are the most important sources of income for slum tenants. Employment opportunities for women, especially the poor and uneducated, are very restricted, however, reflecting the severe social control over women in Bangladesh.[2] Only 22 per cent of women are employed, mostly in occupations derived from their traditional domestic responsibilities: thus 48 per cent of female workers are domestic servants, 25 per cent home-based workers on piece rates, 12 per cent petty traders also based at home, and 7 per cent private tutors (all figures in this chapter are based on research undertaken between 1984 and 1987).

Increasingly, women were forced into black market trading of Indian saris in order to survive. Limited employment opportunities for women, and lower wages compared with men's, means that the pressures on poor households dependent on female labour for survival are particularly severe. In contrast, the labour market for men is much more diverse, and opportunities for work greater. Even among the very poor, income varies greatly. Four groups of tenant households can be distinguished:

- "rich" traders and workers with regular jobs (7 per cent of tenant households) with an average monthly income three times the SPL, and able to invest in productive assets
- medium and small traders (29 per cent) with an average income slightly above the SPL
- labouring households with male workers (38 per cent), with few assets and with debts, and incomes around the SPL
- labouring households dependent on female labour (26 per cent): these are the most vulnerable, with few assets, heavily in debt, and incomes averaging only three-quarters of the SPL.

These latter households are headed by women, or by men who are too sick to work. In either case they rely on women's income, and as a result there has been a nutritional crisis in almost all these households. Seventy per cent of this group are moderately to severely malnourished, and 22 per cent of children under five in this group are severely malnourished, accounting for 68 per cent of all severe child malnutrition in the slum.

The two profiles that follow show both what leads to economic and nutritional crisis, and how poor slum women struggle to contend with their extreme deprivation. The working conditions and income of Mina in the first profile are similar to those experienced by almost half of all women workers, the domestic servants. The story of Hasna in the second profile illustrates the serious risks faced by women who have to resort to illegal marketing in order to survive.

Mina, a Hindu widow

In 1984 Mina, a Hindu widow aged about forty, lived with her son aged twelve and her three daughters aged ten, seven and two. All except the son were severely malnourished. Mina herself weighed only 25 kilogrammes and suffered from a chronic "gastric ulcer" for which she was unable to afford medical treatment or relief. There were two income earners for three dependants. Mina worked ninety-one hours a week as a domestic servant and water carrier, and her son thirty-five hours as a piece-rate bread seller. Household income in cash and kind was around 400 taka per month which placed it in the poorest 10 per cent of households in the slum. Mina owned no productive assets and minimal household assets.

Mina was born in a village in Khulna District. Her paternal grandfather was a farmer and owned around 1.3 acres of land. Her father was disinherited as a result of protracted family quarrels and migrated to Khulna when Mina, the youngest of four daughters, was about two years old. He worked as an unskilled labourer, and Mina's mother as a domestic servant. Like her older sisters, as soon as she was old enough Mina worked as a servant. At the age of twelve her marriage was arranged to a distant cousin, Ashok Lal. On her marriage Mina migrated to her husband's home village where he farmed around 3 acres of land jointly owned with two brothers, and also worked as a potter.

With the Pakistani onslaught on Bangladesh in 1971, because they were Hindu, Mina's husband sold the family land for much less than it was worth. They were threatened and robbed, and finally fled to a refugee camp in India. They returned after liberation in 1972 to find their house occupied by Muslims. The Hindu potter with whom Ashok Lal had worked had also fled. Without any means of livelihood they went to Medja Para slum in Khulna where Mina's parents lived. Ashok Lal worked as a vegetable seller with Mina's father, and Mina again became a domestic servant in the house of one of the slum landlords.

By 1979, it was apparent that Mina's husband had contracted TB. They moved to a small plot of land adjacent to Mina's employer as his tenants, and her husband sharecropped a small plot for the landlord until his illness compelled him to stop. He suffered for a further two years until his death in May 1984, six months before we first visited Mina.

During her husband's illness and in the first six months after his death, Mina was totally dependent upon the patronage of her employer/landlord. He arranged for her husband's funeral, and for hospitalization and treatment for Mina during a severe attack of dysentery. He also arranged an interest-free loan for the marriage of an elder daughter just before her husband's death, and occasionally donated food to the family when the children were very hungry. Mina's relationship with her patron was deferential. Without his charity she might not have survived at all, but she realized that he paid her less than 50 per cent of the going rate for domestic servants.

In December 1986, Mina reported that support by her patron had gradually diminished. She still worked for him for the same wages as two years earlier. Her son had managed to secure an apprenticeship with a Hindu workshop owner through the charity of another neighbour. The three meals a day he received as payment in kind ensured that his nutritional condition remained adequate for his height. The three unmarried daughters, however, were forced to beg together in the market-place. Their nutritional condition remained critical and all three had contracted TB. Significantly, Mina had refused an offer of free medical treatment for them, saying that their fate was in the hands of the gods. Neighbours, however, felt that Mina's own ill-health and severe malnutrition, together with the economic pressures of maintaining and marrying the daughters, were important factors contributing to her decision.

Hasna, an illegal trader

In 1984 Hasna was thirty-two and lived with her husband Abdullah, aged about fifty, their four daughters aged fifteen, twelve, seven and twenty-one months and their niece aged seventeen. Hasna and the youngest daughter were severely malnourished. The household was under great economic strain. Abdullah who worked as a fisherman was chronically ill with various unidentified complaints and could not work regularly. The twelve-year-old daughter worked as a servant in the main market. Income in cash and kind was around 600 taka per month (among the poorest 25 per cent of households in the slum).

Abdullah Rahman was born in a village in Khulna District. His paternal grandfather was once a large farmer whose land was gradually washed away by the river Sheshbati; his father then worked as a woodcutter. Abdullah was the third of his parents' six surviving children. He was sent at the age of eleven to his uncle in Khulna to work as an unskilled labourer, sending back his wages to his family. Six or seven years later his father died from protracted diarrhoea at a time of crop failure and hardship in the village. The family migrated to kin in India, where they remained for four or five years, returning once the food emergency in the village had ended.

At the age of twenty-seven or twenty-eight Abdullah married Hasna who was about ten years old. Two daughters were born. In 1968 Abdullah became partially paralysed and they went to Khulna in hope of treatment. They lived in a squatter settlement and survived for the first year by Hasna begging in the market with her youngest daughter, who subsequently died. During this time Abdullah was treated by a homeopath. His paralysis subsided to an extent which enabled him to start working again in the wood depot of a man who had given alms regularly to Hasna.

During the upheaval of 1971, Hasna's family went to Abdullah's home village for safety. When they returned to Khulna in 1972, Abdullah's employer had been killed and his depot looted. After a variety of casual jobs, in 1973/4 Abdullah started dealing in the lucrative black market for wheat, and gradually accumulated a working capital of 10,000 taka. In 1976/7 there was a dramatic decline in the black market supply of wheat, and Abdullah was forced to disinvest. He spent 5–6,000 taka of his capital on consumption and on medical treatment for Hasna, who was severely ill with her seventh pregnancy and later miscarried. In addition 3,000 taka were stolen from his home.

Since then they have again been struggling to survive. Abdullah was forced to close his wheat business and did a variety of casual jobs until he started fishing, but his worsening health meant that he could not work regularly. Hasna also started work as a domestic servant, but stopped on the birth of her youngest daughter in 1982; her third daughter (aged ten) took her place.

In May 1984 the household was under severe economic strain. Income in that month was less than expenditure, despite attempts to cut back on the number of meals consumed. In the short run, the family survived by spending the last of a 1,000 taka consumption loan. In July Hasna went into the black market in Indian saris which despite its risks was becoming increasingly common among women in the slum. The market is highly differentiated. There are large and small traders, middle-women and piece-rate sellers, all with different levels of income and risk. Most slum women were engaged in work at the lower end of the scale on a piece-rate basis, earning between 2 and 5 taka per sari.

Hasna started as a small trader by borrowing 1,000 taka from three women in the slum at an interest rate of 10 per cent per month. She went to the Indian border three or four times weekly and bought three to six saris at a time from an Indian woman. Purchase and sales prices vary with the quality of the saris. The profit margin was between 20 and 50 taka per piece. Indeed the *potential* earnings of Hasna's activity were by far the highest of any available to women in the slum.

By November 1984, Abdullah's health had deteriorated further and he was only able to work for six days a month. Hasna's niece had joined in her trading, and income in cash and kind was around 1,000 taka per month. The economic situation had clearly improved but Hasna understood the risks of illegal marketing. Four of her women friends had been imprisoned; Hasna herself had been severely beaten and had 750 takas worth of saris confiscated at the border in October, representing all her savings. Female illegal sari traders in the slum revealed that sexual harassment and favours were also common. A girl aged about thirteen from one of the poorest families in the slum had "disappeared" at the border during a trip with the women to buy saris. They were convinced that she had been taken for prostitution to Calcutta.

In December 1986 Hasna's family were once again in an economic crisis. Abdullah could not work at all due to ill-health. New legislation by the Bangladesh government made it illegal not only to smuggle Indian saris over the border but also to sell them within Bangladesh. This resulted in tighter controls at the border, and Hasna had been unable to trade for the previous two months. Her last attempt resulted in the confiscation of 4,000 takas worth of saris and a severe physical beating by border officials.

From October to December the household's income had plunged to around 30 per cent of the SPL, and consumption loans had increased to 900 per cent of monthly income. All the family was malnourished and Hasna was pregnant for the ninth time, clearly not by her husband, and she did not want to talk about it. Her women friends told us that the tighter border controls had resulted in greater sexual harassment. Whether Hasna had been forced into sexual favours to protect her income, had been raped, or had entered prostitution (increasingly linked with the sari trade) was not clear.

Conclusions

Bangladesh is still a predominantly agrarian society. Only 18 per cent of the population live in urban areas, but urban growth is amongst the highest in the world. Intensification of rural poverty has resulted in increased urban migration by landless households, and has also undermined the ability of the patriarchal family system to maintain women dependants. This has led to an increase in divorced and abandoned women, an increasing necessity for women to work to maintain themselves and their children, and consequently more independent urban migration by rural women.

The stories of Mina and Hasna illustrate the extreme economic and social conditions that define the struggle for survival of millions of urban women. Providing the means for them to earn an adequate living for themselves and their families requires more radical measures than the creation of a female working class for world market production (the garment industry in Dhaka, the export fish industry in Khulna), or the limited income-generating credit schemes aimed at slum women by the government.

Notes

1. The slum food poverty line (SPL) was based on a weighted basket of the food most commonly purchased and consumed by slum households in 1986. It comprised 2,800 Kcals per adult male equivalent consumption unit and cost 319 taka/CU/month in December 1986. There were approximately 30 taka to $1US at this time.
2. See Naila Kabeer, "Subordination and struggle: women in Bangladesh", *New Left Review*, no. 168 (1988), pp. 95–121; M. Chen, "Poverty, gender and work in Bangladesh", *Economic and Political Weekly*, vol. XXI, no. 5 (1986), pp. 217–22.

12 Poverty, Purdah and Women's Survival Strategies in Rural Bangladesh

Naila Kabeer

Poverty enumeration in Bangladesh: an exercise in information management

The enumeration and analysis of poverty in Bangladesh, as elsewhere, is invariably conducted by those who have never had to experience it themselves: policy makers, official advisers, statistical experts and foreign funders. Yet the concepts and measures of poverty generally employed by these groups are not designed to enlighten them about the meaning and causes of poverty. The purpose of poverty monitoring appears to be far less ambitious: to provide various interested parties with highly aggregated statistics on income, consumption, landlessness and mortality rates, to quantify the "unproductive" section of the population who are going to require poverty alleviation measures. The complex problems of hunger, deprivation and insecurity are thus compressed into a few, easily manageable figures which tell the enumerators all they wish to know about potential impediments to economic progress.

Despite the appearance of objectivity generally associated with "number-crunching", counting the poor is seldom a neutral exercise. Rather, it is a process of information management which reflects the needs and priorities of those who control the apparatus of data collection. How poverty is measured, against what standards and in comparison to whom people are called "poor": these issues together constitute the political subtext of poverty measurement. In fact, aggregated statistics on poverty are the result of (and help to perpetuate) politically dominant views of the poor as passive and dependent – "a kind of statistical cannon fodder"[1] for those who plan on their behalf. They sanitize poverty, stripping it of its messy and conflictual dimensions and presenting it in a form that is convenient and unthreatening to policy makers. There is little recognition given to the presence within this "amorphous and undifferentiated mass"[2] of heterogeneous groups of people – women, men and children – who have differing access to the means of subsistence, differing ways of surviving in the face of need and pose differing challenges to achieving economic growth with equity.

In this chapter, I challenge these passive stereotypes by focusing on a sub-group of the poor: poor women, who are subject to a twofold process of stereotyping. Like the rest of the poor in Bangladesh, poor women are regarded as fatalistic and passive, objects of policy making and targets for assistance. Reinforcing this representation of the poor, however, is a culturally constructed one of women as a vulnerable group, dependent on male protection and male support, incapable of doing anything but producing children, i.e. "Poor, powerless and pregnant" (as the title of a 1988 Population Crisis Committee briefing paper would have it).

But poor women are far from passive. In this chapter, I examine the strategies employed by rural women to survive in conditions of poverty and subordination. I hope thereby to demonstrate that, while social norms and practices in Bangladesh construct *all* women as passive and dependent, the extent to which they actively seek to conform to this ideal will depend largely on the options and pay-offs available to them. Both women and men are engaged in daily strategies of survival; these strategies are by no means identical since the interplay of class and gender relations gives rise to highly differentiated patterns of constraints and possibilities across the social spectrum. Changes in the strategies poor women adopt to maintain their access to food suggest that they are responding to increasing impoverishment and that the cultural ideal of the passive female is fast becoming untenable, particularly among the poor. These are important changes with long-term implications for the social position of women in Bangladesh, but they remain hidden in aggregated statistical approaches to poverty.

Purdah, patriarchy . . .

Impoverishment is a gendered process in Bangladesh; that is, it is experienced in different ways by women and men, and has different outcomes for them. Both the family-based household, which remains an important locus of subsistence activities, and the market where the poor are driven to sell their labour, offer different sets of options for women and men, and consequently generate different survival strategies.

Purdah, or the practice of female seclusion, prescribes a marked gender segregation in rural tasks and activities, roughly corresponding to an "inside/outside" (*ghare/baire*) divide. It constrains women's ability to move freely in the "outside" world (the fields, the roads and the market-place) and confines them to tasks and activities which can be performed within the precincts of the homestead. In the agricultural process, for instance, field-based stages of production have to be carried out by men; those activities located in or near the homestead are the preserve of women, along with various domestic chores. In the case of rice, the primary food crop, all tasks from sowing to threshing are the responsibility of men. Threshing appears to mark a transition point where women and/or men can participate; all post-threshing stages necessary to turn paddy into rice ready for consumption or sale are supposed to be performed by women. Transport of produce to the *hat* or market-place, on the other hand, is undertaken by men. This division of labour, with women appearing at home-based stages in the sequence of crop production, permits the appropriation of women's labour at the point of sale by male members of the family.

Purdah also operates at the ideological level. It represents cultural ideals about sexual behaviour, family status and female propriety. Families signal their status within the community by their ability to provide the symbolic shelter of purdah to their women, protecting their virtue and moral reputation. Women in turn invoke the constraints of purdah and propriety as the basis of their claims to shelter and support from male guardians. Those who are forced to move outside the boundaries of the homestead generally do so with reluctance, because of the antagonism they encounter and the anxiety they experience. It is this close meshing together of the ideological and material which makes purdah such a powerful controlling mechanism on the behaviour of all women, regardless of class.

The primary domestic grouping in Bangladesh is the family-based household, sometimes composed of extended families, but more frequently, particularly among the poor, of nuclear or even sub-nuclear units. Purdah norms, property rights and familial hierarchies coalesce within the household to produce a corporately organized, patriarchal collectivity. Men tend to control most of the household's material resources, including the labour of female and junior members of their households, and also to mediate women's relations with the non-familial world. Women are socially constructed as passive and vulnerable, dependent on male protection and provision for their survival. They are generally reluctant to seek incomes outside the socially sanctioned relationships of family and kin, first because there are few options to do so and second, because they could forfeit the support of their kin. Women's well-being therefore tends to be tied to the prosperity of the household collectivity and their long-term interests best served by subordinating their own needs to those of the dominant male members of the household.

The asymmetrical relationships which define the terms of interaction between women and men constitute what Kandiyoti has called the "patriarchal bargain" in the specific context of rural Bangladesh.[3] Since women have very constrained access to material resources outside the familial domain, it is in their interests to try to maximize their security within kinship networks. This involves submission to the authority of senior male kin, bearing male heirs so as to secure their position within the conjugal relationship, deferential behaviour and conflict-avoidance, especially with husbands' kin. They also include winning the loyalty of sons so that they remain their mothers' allies, even after their marriages. The lower nutritional status observed for women and girls in all classes of rural household makes sense in this context.[4]

Women give preference to their husbands and sons in the distribution of household food; young girls learn very early in their lives to recognize and accept the prior claims of their brothers and father on household resources.

. . . and women's survival strategies

However, the concept of the patriarchal bargain suggests that, while the terms of interaction between the sexes may be unequal, they are also fluid and capable of renegotiation under differing circumstances. This is evident in the changing strategies by which women and men try to assure their daily survival in the face of increasing scarcity.[5]

Poverty and landlessness modify, but do not override, the influence of purdah in women's lives. Women attempt to cope with poverty as far as possible in ways that will not threaten their kin-based networks and their family's standing within the village community. Often this entails retaining their status as financial dependants, even when male support is precarious and unreliable. However, female dependence is premised on the cohesiveness of the family unit, particularly the conjugal unit since it is the one in which they are expected to spend a major part of their lives. If the family economy begins to disintegrate, dependence becomes increasingly difficult to sustain and women are forced to adopt strategies which represent a break with their former dependent status. However, as long as male support is forthcoming, women's survival strategies will adhere as far as possible to social convention so as not to jeopardize their claims on male family members. There are three different kinds of survival strategies which can be pursued by women without directly infringing the boundaries of purdah: income-replacing activities, intensified self-employment and disguised wage employment.

These survival strategies generally form part of the household's collective efforts to secure its own daily reproduction. Since poverty in rural Bangladesh is causally associated with the absence of assets, the poor rely critically on the productive use of their primary and frequently sole resource, labour power. Men use their labour power in a variety of ways, but generally in market-orientated activities. They offer themselves for wage labour in agricultural production or rural industry; they undertake small-scale trading activities, moving between different kinds of markets to make a profit, or they engage in petty commodity production with household labour. Women also dispose of their labour power in a variety of other ways, but within the narrower parameters imposed by gender segregation in the division of labour. Women's labour contribution is of particular significance among low-income households who may not know where their next meal is coming from. Such households rely critically on women's ability to stretch out their meagre resources through a range of productive activities which require little material investment but make intensive use of informal networks and skills.

Women, often assisted by their children, will glean the rice fields after the harvest or recover rice stolen and stored by field mice. They also gather edible wild plants from the roadside or common ecological reserves. Leaves from various plants like jute and water hyacinth, edible roots like *kochu* and the fleshy stem of the banana plant are all forms of food which used to constitute "famine foods" but are becoming part of the daily diet of the poor. It is also generally women who borrow or beg quantities of rice, lentils or the discarded parts of vegetables from wealthier neighbours. Indeed, poor women invest a great deal of effort in nurturing these neighbourhood networks because they constitute an important resource to fall back on in times of need, in spite of the unequal class relationships expressed in them.

A second set of survival strategies in response to deepening poverty is to draw on those members – women and children – who may not have been required to contribute to household income before. This need not violate purdah since many forms of income earning can be carried out by women within homestead precincts. Growing fruits and vegetables on small homestead plots and rearing poultry and small livestock are common examples of work undertaken by poorer women to earn income. "Share-rearing" of livestock allows poor women to transform their labour power into a productive asset: in return for rearing the animals of wealthier households poor women share the produce, alternate offspring in the case of female stock or half the profit from the sale of male stock.

In addition to various forms of self-employment within the homestead, purdah norms can still be preserved when income is earned on putting-out basis. In one district visited, women were supplied by a local sub-contractor with palm leaves which they wove into mats, returning one mat in every two to the sub-contractor and keeping the other to sell themselves. Similar arrangements were observed for a number of other craft products (quilts, jute goods, etc.). Finally, the performance of domestic chores and post-harvest processing of rice in the homes of wealthier landowners in exchange for payment in cash or kind (usually meals) is another means by which poor women earn money without appearing to violate purdah norms. The waged basis of these different earning activities is disguised by the casual and personalized relationships within which they are conducted, by irregular and often non-cash modes of payment and by their location within the home rather than in a clearly defined work space. Disguised forms of waged work permit women to sell their labour in informal labour markets, but to retain the appearance of seclusion.

Renegotiating gender divisions

However, major changes in the rural economy over the past decades have made it progressively more difficult for purdah to be maintained among poor households. They have led to the renegotiation of aspects of gender relations, including the division of labour, purdah and female mobility, family structures and the security of the conjugal unit. Not all renegotiations of gender divisions favour poor women or are welcomed by them. Poor women attempt to minimize the adverse consequences of change, often by sacrificing the emancipatory potential it may also contain. Their coping strategies lead to the merging of male and female incomes, clandestine forms of economic activity and bargaining over asset holdings (including land and livestock). In some circumstances, patriarchal protection is no longer forthcoming, and strategems and bargaining become redundant.

While growing impoverishment and landlessness have increased the number of women seeking income-earning opportunities, important sources of employment for rural women are being eroded by the mechanization of post-harvest crop processing, net weaving and oil crushing. The imperatives of survival are leading to a perceptible restructuring of the gender division of labour and the increasing presence of women in areas of work formerly closed to them. Some changes are evident within the agricultural process itself. Hitherto strictly enforced rules preventing women engaging in field-based stages of rice production are showing signs of crumbling and women are being employed in harvesting, weeding and transplanting work. In addition, they have entered in large numbers into public rural works projects, into small workshops and mills, and into petty trading activities in the bazaar economy.

These activities are generally taken to represent an unequivocal break with the boundaries permissible by purdah and a challenge to cultural ideals of female dependence. But they will not necessarily translate into autonomous survival strategies for women, if disposition of female wages remains within the domain of patriarchal decision-making processes. Male and female earnings are often treated as complementary elements in the household's collective strategy for survival. In one family interviewed, for instance, daily consumption needs were met by a male breadwinner who worked as an agricultural wage labourer; the female income, earned though intermittent processing and sale of small amounts of purchased paddy, was saved for emergency needs. In another household, where female members engaged in share-rearing chickens, daily needs were partially met through the consumption and sale of eggs while the poultry represented an easily convertible hedge against crisis. Income-pooling strategies, however, generally curtailed women's financial autonomy through customary rules about family decisions.[6] Personal accumulation and personal consumption (for example, cigarette purchases and tea drinking at the village store) are sanctioned for men, but not for women. Women's claims to familial resources thus continue to be subordinated to those of men, regardless of the changing significance of their contributions.

However, most women are aware of the precarious nature of their claims on male guardians and breadwinners. The obligations associated with marriage and family relations are difficult to sustain when poverty has begun to erode their material basis. Male survival strategies in the face of increasing impoverishment often entail the abdication of responsibilities to wives and dependants; a number of studies confirm the link between poverty and the incidence of desertion, divorce and female-headed households. Development workers in the field also point to marked seasonalities in the incidence of divorce (men are most likely to abandon their families in the hungriest months of the year) and to its association with the aftermath of crisis events, such as floods and famine.

One strategy by which women seek to cope with the insecurities of male protection is by carving out small areas of relative autonomy which may not appear to dispute the terms of the patriarchal bargain, but in reality attempt to renegotiate its basis. Efforts by women to gain a measure of autonomy within the domestic hierarchy frequently entail clandestine forms of behaviour. One example is the long-standing practice of secret saving and lending of money or rice through women's neighbourhood networks. Another example is the reliance on trusted male or female relatives to assist in various covert forms of earning. Thus one woman kept her membership of a local development project secret from her husband. She used loans from the project to finance and profit from a small business run by her sister's son.

Other strategies do not rely so obviously on concealment, but revolve around the choices and trade-offs made by women in negotiating over household asset holdings. Women from families with some property generally choose to trade off dependence on brothers against dependence on husbands by waiving their rights to any share of parental property in exchange for their brothers' support in the event of marital crisis. This option is obviously not open to the majority of poor women who come from landless households, but implicit, gender-constrained trade-offs are also evident in decisions concerning other forms of asset holdings. These can be illustrated by looking at the acquisition and disposal of livestock by poor households.

The arrangements for share-rearing livestock and poultry have implications for the degree of control women exercise over its benefits. Women may, for example, prefer to sell their stock to a middleman who comes to their homestead, even though he gives a lower price than the bazaar, because this arrangement gives them greater control over the income.[7] Relying on a husband (or other male intermediaries) to procure a higher price in the bazaar carries the risk that they will not receive the full returns for their labour. Alternatively, a husband can use the information he acquires on his wife's income through selling her livestock to pressure her into spending it according to his order of priorities. There are also gender implications in the choice of animals acquired for rearing purposes. The bathing and feeding of larger animals, like cows, bullocks and buffaloes, requires the participation of adult males. Such livestock is, in principle, owned collectively by the family as a whole, but in practice is usually appropriated by male members. Smaller stock (sheep or goats) and poultry can be tended entirely by women and children, and are therefore more likely to be seen as women's property. Women often express a preference for share-tending smaller animals, partly because they are cheaper and easier to rear, but also because they represent an independent asset in the event of marital crisis.

In their survival strategies, poor women combine elements of autonomy with dependence on their male partners because the patriarchal bargain is unstable in conditions of poverty. Without wishing to precipitate the collapse of the bargain, they nevertheless seek fall-back livelihoods and savings, should male protection and provision cease. The rise in poverty-related divorces and desertions, the emergence of large numbers of female-headed households, the presence of a female labour force outside culturally prescribed boundaries, all reflect the unwillingness or inability of men to discharge their customary responsibilities towards their dependants – wives, sisters, mothers and children. Survival strategies in situations where women are sole breadwinners generally display a diminished concern with propriety and purdah; scarcity becomes the overriding issue. Returns to labour rather than status considerations determine choice of employment, and women will take work wherever they can find it: by the roadside, in the fields, in bazaars and in brothels.

The daily struggle for survival in these situations leads to an enforced rejection of the notions of feminity and modesty conventionally associated with respectable womanhood; poor women learn to develop sharp tongues and aggressive demeanours to deal with the hostility they encounter when they encroach on male space. There is also evidence that women are increasingly migrating to other districts or to cities, either with their dependants or by themselves, in search of economic opportunities.[8] (Pryer's chapter in this volume contains a discussion of how women cope in the very different circumstances of urban poverty.) This new evidence is in sharp contrast to previous decades when migration was primarily male. It further highlights the poverty-induced changes which are occurring in purdah constraints and female mobility. Finally, there may come a point in the downward spiral into poverty when, like men, able-bodied women are forced by economic need to abandon their responsibilities to dependants. Some women place their children in orphanages to be fed by local charity; others simply leave their children or ageing relatives to fend for themselves in the informal bazaar economy.

From survival to empowerment: collective responses to poverty

The elements of "autonomy" in women's survival strategies are small areas of freedom from personalized forms of male control rather than freedom from patriarchal controls in general or from the constraints imposed by class relations. Female-headed households, while free from male authority, are among the poorest strata in rural society precisely because of the very definite limits imposed on women's capacity to secure their own survival, independent of male intermediaries. The survival strategies I have described are simply that: attempts by the poor to reproduce the basic conditions of their daily existence.

Such attempts seldom form a springboard for economic accumulation, nor do they challenge the basis of patriarchal power. Women's subordination in Bangladesh is produced by interlocking and asymmetrical relations of marriage, property and production, reinforced and perpetuated by powerful ideologies of gender inequality. It is deeply entrenched in social structures and social consciousness, and likely to remain intact as long as challenges to it are mounted in a piecemeal and fragmented fashion. It is the *individualized* nature of the survival strategies discussed so far which renders them incapable of transforming the position of poor women in the interlocking hierarchies of class and gender.

More collective responses to poverty are, however, being organized by a number of progressive grassroots development organizations in Bangladesh who seek to combat poverty through a process of "empowerment", by which people enhance their capacity to analyse, understand and act on their situations; by which they are enabled to locate the causes of their oppression as existing outside, rather than within, themselves and to take up the struggle to change it. It requires the poor to recognize that they are an oppressed group, but first it requires them to recognize that they are oppressed individuals.[9] The process of empowerment thus begins, but does not end, with the individual.

While there are important differences in the approaches of various NGOs to the processes by which empowerment is achieved, there are also important similarities. One important difference lies in how they perceive the provision of services to the poor: as a means of creating organizational networks or as a seedbed for new forms of dependence. An important similarity, on the other hand, is the significance attached by these NGOs to collective education as a means of tackling the ideological entrenchment of oppression. For poor women, the discovery that their poverty, isolation and sexual subordination are not divinely ordained or biologically determined (or for that matter individual to them) is a powerful one. It has led to their active participation in a number of attempts to mobilize by the organizations of the poor. Women and men have acted in solidarity to assert their rights to unutilized government land, to trading rights, fair wages and equality before the law. In addition, women's groups have challenged, sometimes with male support, practices which underpin women's subordination such as dowry, polygamy and wife-beating. By breaking the silence which has in the past provided a *de facto* legitimation for these forms of subordination, poor women are moving from merely struggling to survive within the parameters of class and gender to a position of challenging the parameters themselves.

Notes

1. Beck, Tony, "Survival strategies and power among the poorest in a West Bengal village", *IDS Bulletin*, vol. 20, no. 2 (Lewes, 1989), p. 24.
2. Chambers, Robert, "Editorial introduction: vulnerability, coping and policy", *IDS Bulletin*, vol. 20, no. 2 (1989), p. 1.
3. Kandiyoti, Deniz, "Bargaining with patriarchy", *Gender and Society*, vol. 2., no. 3 (1988), p. 275.
4. Kabeer, Naila, *Monitoring Poverty as if Gender Mattered; a Methodology for Rural Bangladesh*, IDS Discussion Paper 255 (Brighton: Institute of Development Studies).
5. This section is based mainly on interviews conducted by the author in 1987–8 with poor women from different rural districts of Bangladesh.
6. Standing, Hilary, "Resources, wages and power: the impact of women's employment on the urban Bengali household", in Halsh Afshar (ed.), *Women, Work and Ideology in the Third World* (London: Tavistock Press, 1985).

7. Akhter, Farida, and Fazila Banu Lily, *Women's Role in Livestock Production in Bangladesh: an Empirical Investigation* (Dhaka: Bangladesh Agricultural Research Council, 1984).
8. Rahman, Atiq, Simeen Mahmud and Trina Haque, *A Critical Review of the Poverty Situation in Bangladesh in the Eighties, Vol. 1*, (Dhaka: Bangladesh Institute of Development Studies, 1988).
9. Freire, Paulo, *Pedagogy of the Oppressed* (Harmondsworth: Penguin Books, 1972), p. 141.

13 Peasants under Contract: Agro-food Complexes in the Third World

Michael Watts

Farming contracts link "independent family farmers" with a central processing, export or purchasing unit which regulates price, production practices and credit arranged in advance under contract, thereby replacing open-market exchange. In the United States, and across much of Western Europe and Japan, contract production has emerged as a centrepiece of contemporary agro-food complexes. Contracting or *vertical co-ordination* in US agriculture now far exceeds *vertically integrated* agribusinesses (corporate production) in terms of gross output. By 1980 one-third of US farm output by value was produced under some form of contract.

Contract farming represents a form of social organization of growing significance, in which plants and animals are produced on land in relation to the complex and changing profit conditions of global capitalism. It does not ensure the preservation of family farming (small-scale commodity production), but rather is the means to introduce distinctive work routines, new on-farm technologies and labour processes, a further concentration and centralization of capital in agro-food systems, and not least to deepen the form of appropriation by which rural production processes (farm inputs and services) are converted into industrial products by agro-industrial, capitals and subsequently reincorporated into agriculture (see Buttel's chapter in this volume). Contract farming marks a critical transformation and recomposition of the family farm sector as capital saturates the entire agro-industrial complex without directly taking hold of production.

Contract farming on a global scale

Contract production is no longer confined to advanced capitalism and Euro-American agro-food complexes but operates through global circuits of capital as part of international commodity markets.[1] In the Third World contract farming has a dual origin. First, contracting competes with and partially replaces plantation agriculture employing free wage labour, as foreign agribusiness is subject to nationalist pressures, threats of expropriation and local regulation, and new conditions of profitability within a changing international division of labour. In some cases plantation and estate owners may retain a foothold in agriculture by contracting for export – for example United Fruit's banana operations in Honduras. Conversely, foreign agribusiness may terminate production quite abruptly. California-based Bud Antle liquidated its horticultural operations in Senegal in 1976 and was promptly replaced by a state-run firm (SEMPRIM) contracting fresh vegetables (beans, tomatoes, melons), primarily with women growers, for EEC markets. Both cases suggest contracting emerges with the decomposition of plantation/estate production linked to a persistence of the classical export commodities (sugar, tea and palm oil) now cultivated under contract by a variety of local growers.

Second, independent peasants, and sometimes newly settled pioneer farmers, are drawn into new corporate forms of social organization under state and/or private auspices producing a variety of commodities for domestic consumption and export. In Mexico for example, Pillsbury, Campbell Soup, General Foods and other US transnationals engage in contract farming for strawberries, tobacco, tomatoes and cocoa, while state corporations contract heavily for sorghum and soy. Contracting appears in this case as the recomposition of peasant producers; independent growers are institutionally captured by, and socially integrated into, new production complexes.

The growing popularity of contracting is not unrelated to the hegemony of IMF austerity measures in a debt-ridden Third World, and the desperate search for private-sector mechanisms to revive flagging export sectors and alleviate foreign exchange shortages. In the 1980s US agribusiness and USAID have actively promoted contract farming as a "dynamic partnership" between small farmers and private capital which promises market integration, economic growth and technical innovation while protecting the rights and autonomy of the grower via the contract. The World Bank in its influential assessment of the agricultural crisis in sub-Saharan Africa – the so-called Berg Report – targets contracting as an exemplary means to enhance the private sector. Similarly, the International Finance Corporation (a division of the World Bank) and the British Commonwealth Development Corporation (CDC)[2] have pioneered palm oil, cocoa and rubber contracting – in their vocabulary core-satellite, nucleus-estate or outgrower schemes – across Asia, Africa and Latin America. These are advertised as models of technology transfer, small-farmer settlement, export-led growth and integrated rural development (i.e. nurturing a conservative "middle-class" peasantry).

Contracting has a particular appeal, of course, for Third World ruling classes and state managers imperilled by liquidity crises and sensitive to the populist rhetoric of targeting the rural poor, promoting smallholder development and "putting peasants first". Some of the largest outgrower contracting schemes – tea production in Kenya, tobacco and livestock in Thailand, rubber in Malaysia, palm oil in the Philippines – are public-sector enterprises in which the state is typically the dominant partner in joint ventures with transnational agro-industry and foreign banks.

Contract farming is reshaping the face of Third World agriculture and marks a new phase in the integration and subordination of peasantries to the world market. The export-orientated horticultural industry in Central America, Southern Europe and parts of North and Eastern Africa has pioneered the internationalization of contracting, advancing what Maureen Mackintosh calls the erosion of market relations between growers and exporters, freezers, canners and brokers.[3] Kenya, a model of plantation and estate production under British colonial rule, now sustains the most extensive production contracting in sub-Saharan Africa: 12 per cent of all peasant farmers grow sugar, tea, tobacco and fresh fruit and vegetables under contract to state and private agro-industrial contractors, producing 17 per cent of total farm output and 30 per cent of total marketed output. Some 100 Kenyan horticultural buyer-processors export 31,000 tons of fresh produce, contracting an astonishing array of commodities: carnations for London florists, speciality vegetables for the British Asian Community, and green beans for Parisian winter markets. If contract farming is the system of the future for Third World peasantries, does it promise growth with equity or something closer to debt-peonage? Is contracting acting to further differentiate peasantries, and in some cases to exclude the rural poor altogether? Does contracting represent a form of flexible accumulation integrating peasants into new networks of control and subordination? How and in what ways can peasants under contract resist and organize?

The social organization of contract farming: a peasant strategy?

Contracting represents a major front along which capitalism advances into the sphere of family or household production. It employs a variety of social and organizational forms, and is not only aimed at peasants. Nor is it the monopoly of agribusiness alone but is increasingly a state strategy, often in alliance with local and foreign capital. Contract relations may systematically exclude large segments of the rural poor and target middle peasants or local capitalist farmers. Three patterns of contract farming appear in the Third World:

- large, centralized and frequently state-owned nucleus-estate schemes with thousands of peasant outgrowers and a central processing unit
- contracts primarily with large capitalist growers, whose associations negotiate with state and labour
- small peasant contracting usually with local, and sometimes foreign, merchants or exporters.

The first pattern is associated with classical export commodities (sugar, bananas, tea, cocoa) with important economies of scale in processing, high co-ordination requirements (i.e perishability), and labour-intensive crop maintenance. Typically, such schemes combine peasant outgrowers with estate production and heavy investment in industrial processing facilities. Smallholder growers are selected from well-to-do peasants and are little more than capacity contractors to ensure regular throughput to the processing unit.

The second model is exemplified by United Brands and Standard Fruit in their Central American banana operations. Both companies adopted contract production under sustained state-led nationalist pressure in the 1950s which had erupted in labour militancy on their plantations. Banana production was, however, contracted to capitalist growers ("associate producers") organized into an association of capitalist growers (ANBI) who negotiated with labour and the state. By the 1980s sixteen large contract growers supplied 31 per cent of Honduran bananas while almost two-thirds came from United Brand and Standard Fruit corporate plantations. A mere 8 per cent was supplied by four peasant co-operatives producing under contract to Standard Fruit.[4] In Costa Rica, Del Monte contracts with thirteen growers (each in excess of 250 ha) who account for two-thirds of its banana supply.

A third pattern is small peasant contracting usually with local, and sometimes foreign, merchants and exporters. Njoro Canners is a joint venture between a French company and local Kenyan merchants who contract with 15,000 small growers in

Western Kenya. Three-quarters of those under contract are women who cultivate high-quality green beans on tenth-of-an-acre gardens in one of the most densely settled areas in Africa. Lebanese merchants in Senegal similarly purchase beans, melons and fresh vegetables for supermarkets from small peri-urban growers, many of whom are loosely organized into *groupements* or associations under the direction of a local chief.

Capital and the labour process: the logic of the contract

For neo-classical economics and agribusiness, the contract is the embodiment of market mutuality: freely entered, the contract allows growers to make better use of their specific endowments in imperfect markets and to arrive at combinations of income, effort and risk reflecting their resources and tastes.

Contracts are typically one year in duration but are enormously diverse in content and legal character: *market specification contracts* are future purchase agreements which determine quantity, timing and price of commodities sold; *resource-providing contracts* specify the sorts of crops to be cultivated (via seed provision), some production practices and the quality and standardization of the crop through the provision of technical packages and credit; *production management contracts*, associated with large outgrower and nucleus-estate schemes, directly shape and regulate the production and labour processes of the grower. In the Mphetseni pineapple scheme in Swaziland, owned by Libby's, the production management contract determines *all* on-farm operations in a highly regimented work routine. Field operations are compulsorily performed by Libby's own field labour in the event of grower default, and the company retains the right of eviction if the provisions of the lease are broken.

The contract stands between the open or spot market in which independent producers sell to independent buyers and the vertically integrated agribusiness incorporating all operations from production to consumption in one enterprise. In all forms of contracting with household producers the contractor exploits a peasant "labour market" rather than a class of rural proletarians. The grower provides labour power, land and tools, while the contractor provides inputs and production decisions, and holds title to the product.

In lauding the freedom of the contract, its advocates obscure both the degree of economic compulsion and the power exercised by the contractor. Authoritarian forms of contracting control work conditions in a manner that renders household labour unfree in the sense that it is directly distributed, exploited and retained through political-legal mechanisms. Peasant growers become what Lenin called "propertied proletarians", growing corporate crops on allotments. "We do not know how our canes go in or how the sugar comes out," remarked a Kenyan sugar grower, "we only get money." A woman Tanzanian tea grower put it more succinctly: contract work, she said, is like "the big slavery" (see Mbilinyi's chapter in this volume).

Growth with equity?

The smallholder component in contracting is often little more than a rhetorical device to legitimate corporate investment. TNCs conclude contracts only after a careful selection and screening process which privileges capitalized growers and small-scale agricultural businesses. In the Ubombo Ranches sugar scheme in Swaziland, a joint venture between Lonrho and the Swazi government, two-thirds of outgrower output is supplied by white capitalist farmers and "advanced" Swazi growers using wage labour. Similarly, Nestlé's milk suppliers in Chiapas, Mexico, are stock farmers able to buy at least twenty head of imported cattle and to invest heavily in irrigated lands and capital to finance additional sowing of pasture. In these cases the contract is in effect a sub-contract between capitalist enterprises.

In the classic smallholder schemes such as the African state-run tea and sugar schemes, growers are also markedly differentiated. In the Malawi Tea Authority 10 per cent of tea growers reap 41 per cent of total revenues, and in the Kenya sugar, tea and tobacco parastatals 10–15 per cent of growers accounted for close to half of the output and systematically invested in hired labour and land. Kenya's Mumias sugar scheme employs a landholding threshold to screen growers which of necessity excludes land-poor households, as does the contractual obligation for Kenyan tobacco growers to finance a curing shed for the British American Tobacco Company.

There is thus a marked social division within the outgrower schemes between land-rich capitalist growers who employ wage labour, and a substantial middle peasantry who may earn investible surpluses but who depend largely on household labour. The possibility of accumulation contributes of course to land speculation, the emergence of absentee farmers, growing differentiation among growers (and between regions) and the further marginalization of poor peasants increasingly employed as casual labourers on central estates and on the farms of large growers.

Poor peasants may, and do, participate in contracting, and some of the evidence from African horticultural schemes indicates that in the short term income gained can be significant. Yet the risks for *all* peasant growers are substantial. In a competitive world market horticultural contracting is extremely volatile, subject to wide price swings and overproduction. The boom and bust cycle, and the ebbs and flows of cash through the local peasant economy, make for gross instabilities in income and little economic security in the absence of local markets for contracted produce.

Not least are the consequences of export contracting on local food security. In spite of the existence of contracts which stipulate the proportion of grower holdings to be devoted to subsistence crops (which serve incidentally to lower the reproduction costs of labour), food shortages are endemic and local cereal markets volatile, subject to wild seasonal price swings. A study in Western Kenya shows that substantial increases in income from sugar over a twenty-year period did not substantially improve nutrition of women or pre-school children. Morbidity patterns and malnutrition remained high throughout the sugar belt (50–70 per cent of children and women are sick at any one time), and unstable food prices made for periodic food shortages and inflationary squeezes among the land-poor and landless classes. "Sugar eats everything" is the refrain of Kenyan sugar growers, just as Kenyan tobacco is seen locally as "bitter money" for contract farmers.

Struggles over the contract

If contract production among peasants aims to exploit household labour through dense networks of dependence and subordination, in what sense is a disciplined labour force actually produced? Can contractors manufacture consent among peasants under hierarchical and coercive conditions in which, unlike the forms of capitalist production Marx analysed surplus appropriation is not obscured within the production process? Is the system self-reproducing and self-disciplining?

On one hand, the contract is clearly a means of subordination. The company often retains legal title to crops and inputs and temporary rights in the farmer's land and labour. In the case of debt and systematic failure to comply with contract specifications, the company can evict and dispossess, even though the contract is quite frequently "signed" by illiterate peasants and hence popularly misunderstood. Quality controls in particular are a source of company manipulation as contractors regulate supply through arbitrary tightening of produce standards, and crop quality typically emerges as an arena of enormous conflict and enmity. It is also not unusual for a company to tear up its contracts. For example, ALCOSA, a Guatemalan subsidiary of Hanover Brands/Birdseye, contracted with 2,300 peasants in Chimachoy for cauliflower and broccoli. In 1980 it unilaterally suspended purchases when faced with a severe surplus, leaving angry peasants with no alternative market, massive losses and little in the way of food.

Contractual relations further subordinate growers to buyer-processors through ties of credit which can threaten to transform the independent growers into bonded non-wage labourers. Evidence from small asparagus and barley growers in Peru (supplying breweries) and sorghum outgrowers in Colombia shows how peasants accumulate debt as rapidly as investible surpluses. The seeds of indebtedness often grow in the fertile soil of changing terms of trade between grower and buyer. In a state-run rice outgrower scheme in Cameroon (SEMRY), for example, grower dues (covering interest, production inputs, maintenance costs) increased by 180 per cent between 1973 and 1980 while producer prices rose by barely 40 per cent.

On the other hand, the contract is also a source of tactical resistance. Growers themselves renegotiate and subvert the terms of contracts. Produce may be adulterated (stones added to increase weight), patronage and personal ties are employed to upgrade crops illegally, and inputs intended for the contracted crop are diverted to other farms typically growing subsistence crops. Peasants may also abrogate the contract by selling produce in parallel or spot markets at higher prices. Such leakage prompted Nestlé in Mexico and vegetable contractors in Kenya, experiencing produce losses because of nearby urban markets, to relocate their operations to isolated, "backward" growing regions in which commodity markets were undeveloped.

Systematic leakage of produce erodes company profitability and raises the vexed question of contract enforcement. Legal and property rights are difficult to police and enforce in many Third World settings where local autonomy is strong and juridical apparatuses lack power and sanction. It is not unusual for companies to suspend any faith in formal legal institutions and rely instead on painstakingly constructed relations of trust, patronage and traditional reciprocities – a moral economy of sorts – rather than the word of the law.

In the final analysis, however, the capacity to evict, fine or discipline, the legitimacy of company claims, and the political space of grower resistance are fundamentally shaped by the class and juridical power of the state. The experience of contracting in Africa suggests that land questions are so sensitive, and the legitimacy of states so fragile, that litigation or contractual renegotiation by private or state interests is potentially explosive. Strikes and boycotts by sugar growers in Kenya in 1985 and bloody battles between peasants and the state on Northern Nigerian irrigation schemes suggest possibilities for collective action triggered by a common grower interest in resisting subordination through the contract.

Factories in the fields: contract farming and global Fordism

The deepening of contract production in agriculture bears striking resemblances to so-called post-Fordism or "flexible accumulation" in sectors of industrial capitalism with a growing reliance on multiple outsourcing through tightly specified industrial sub-contracts. Large manufacturing firms may modify vertical integration by deploying dense networks of sub-contractors as a way of sustaining accumulation in the face of heightened competition. Contracting hence represents a deepening of the division of labour external to the firm yet *reintegrating* control. The flexibility conferred by the contract in the most extreme cases can eliminate centralized factory production altogether by the putting out of industrial work to independent households. In the cases of apparel and textiles in Mexico City or Barcelona, contracting to women outworkers "spreads out over the city as if the city were a huge factory".[5] What appears as industrial deconcentration of production is in reality the technical and social integration of dispersed workers subordinated and controlled via credit and patterns of tied contracting.

The emergence of the "world car" and the diffuse factory have their agrarian counterparts in the so-called "world steer", in the corporate strawberries grown by Mexican sharecroppers in California and in the highly standardized Kenyan green beans embellishing Parisian dinner parties. Like his industrial counterpart, the agricultural contractor is able to reduce fixed costs (i.e. land) and to disperse much of the price and/or production risks to the direct producers.[6] Like the putting-out system in textiles, the contractor exercises direct and indirect control over household labour. Peasants work as *de facto* piece workers, often labouring more intensively (i.e. longer hours) and extensively (i.e. using children and other non-paid household labour) to increase output or quality.[7]

Unlike other industrial sectors, however, agriculture has biological and geographical peculiarities. Labour demand is shaped by biological growth processes, and production time exceeds labour time. The capacity for self-exploitation among peasant growers, and the biological basis of agricultural production, suggest that capitalists need not engage directly in production as much as discipline and control peasant labour. In the words of a Kenyan sugar grower: "Since you agree to plant sugar there is a rope around your neck that connects you to the company."

Contracts, capitalism and unfree labour

Contract farming suggests that unfree labour is both an anomaly and a necessity in capitalist development.[8] When peasants are subordinated under contract, capitalism exploits household labour which is in a fundamental sense unfree. The state is naturally pivotal in the enforcement, arbitration, recruitment and indeed often the direct exploitation of this unfree labour. The political and ideological requirements of contracting explain why the state is imperative in maintaining this particular production regime and why contracting is often conducted directly under state auspices. Peasant producers under contract and located in varying positions of unfreedom constitute a distinct class and may be seen as a fraction of an emerging global proletariat.

By the same token the exploitation of household labour power through contracting reveals that capitalism may contribute to the reproduction of non-wage labour simply because free labour in some sectors is not required. In agriculture the use of unfree non-wage labour can fulfil the same function as technology in industrial capitalism, namely as a means to cheapen and subordinate labour or to substitute for free wage labour. And it is entirely possible that contracting is a preferable strategy for capital when free wage workers can threaten its control. Whether the political, legal and ideological conditions required to maintain this contractual form of unfreedom can be maintained is, of course, a question of class capacity and class struggle. What is clear is that the social contradictions within contracting can generate oppositional energies and popular struggles capable of challenging the relations of dominance on which the "free" contract rests.

Notes

1. For an excellent bibliography of contract farming in the Third World, see Diana de Treville, *An Annotated and Comprehensive Bibliography of Contract Farming* (Binghamton, NY: The Institute for Development Anthropology, 1987).
2. The CDC is a British financial institution active in promoting contracting since the 1950s; it currently sponsors 34 outgrower schemes embracing 600,000 smallholder growers of sugar, tea, coffee and palm oil. See *CDC and the Small Farmer* (London: Commonwealth Development Corporation, 1984).
3. Mackintosh, M., "Fruits and vegetables as an international commodity", *Food Policy*, no. 2 (1977), pp. 277–92.
4. This shows clearly how contracting to smallholders may fulfil a largely ideological function for large capital. Texaco, for example,

has made much of its commitment to the small peasant and to Africa's food problems in its Texagri scheme (contracted cassava production for dried meal) in Southern Nigeria. The 141 small peasants involved provide only one-quarter of the dried cassava (*gari*) production, and the entire project is of dubious social and economic benefit (not least for Texaco!). Nigerians believe the *gari* to be unpalatable and in 1986 it was being given away at Texaco gas stations in Nigeria.

5. Casals, M., *L'Economia de Sabadell* (Barcelona: Ajuntament de Sabadell, 1985).
6. The chairman of British American Tobacco Company pointed out that "the dispersal of growing helps protect us from the vagaries of weather" (cited in K. Currie and L. Ray, "On the class location of contract farmers in the Kenyan economy", *Economy and Society*, vol. 15 (1986), p. 473).
7. It has been estimated that in the case of East African tea – like other contracted crops, extremely labour-intensive – the actual return to grower labour (estimated at, on average, 2,000 hours per acre per year) is *less* than the wage paid to estate labourers (who are seen to be among the most exploited of the rural poor).
8. See R. Miles, *Capitalism and Unfree Labour* (London: Tavistock, 1987), especially pp. 181–95.

14 Biotechnology and Agricultural Development in the Third World

Frederick H. Buttel

In 1983 Martin Kenney published an important article in *Monthly Review* entitled "Is biotechnology a blessing for the Third World?"[1] Most of Kenney's observations and arguments – particularly his scepticism about optimistic predictions of the contributions of biotechnology to international development – are still relevant. This paper updates Kenney's analysis by identifying how the Third World is, or is likely to be, disadvantaged in relation to biotechnology and the development process. The key argument is that while biotechnology has considerable potential for making possible relatively equitable Third World development, this potential will be difficult to realize because of structural features of Third World economies within the global economy (also the problem of generalizing across the Third World).

Biotechnology and the Third World: dimensions of dependence and disadvantage

Biotechnology and the "new technological inferiority"

It has long been recognized that scientific and technological disparities are intrinsic to North–South disparities in capital accumulation and living standards, whether as a root cause or a major consequence of Third World underdevelopment. None the less, the ten or so years prior to the rapid commercialization of molecular biology in the early 1980s can be seen in retrospect as an era in which global technological disparities in agriculture were narrowed to a degree unprecedented in this century.

In the 1970s the techniques available to, and the capacity of, public research institutions in several countries of the South came to rival those of many advanced industrial countries. Due to foreign aid, training and research efforts from multilateral institutions such as the UN Food and Agriculture Organization (FAO) and the International Agricultural Research Centres (IARCs), and Third World government investments in personnel training, facilities and research programmes, a significant number of countries (especially India, Brazil, Mexico and China) developed the capability to undertake some basic research and virtually all the types of applied agricultural research then being undertaken in the advanced countries. Tens of others had sufficient capacity to modify and transfer new technologies developed by the IARCs or developed country research institutes. Even some of the poor countries of the so-called Fourth World (e.g. Bangladesh) were able to make use of public-domain technologies developed elsewhere, with a modest amount of technical assistance from more favoured countries and multilateral research institutions.

Three factors have combined to create a new pattern of Third World technological inferiority in the 1980s. The first, of course, is the breakthroughs in a wide range of "New Information Age" technologies, including but not limited to biotechnology. Their common element is their origin in relatively basic or fundamental sciences that are generally highly underdeveloped in the Third World.

Second, the surge of publicly and privately funded R & D in the advanced countries in new information technologies has been driven by interstate rivalry for dominance in "high-technology". Given international technological competition, the advanced countries try to ensure that research results are spread as slowly as possible to potential competitors.

Third, as the long cycle of global stagnation tends to have particularly devastating impacts on the Third World, increasingly it has faced state fiscal crises and externally imposed austerity programmes, leading to decreased public investment in R & D, including agricultural R & D.

The "new technological inferiority" is the fundamental backdrop against which other dimensions of biotechnology dependence and disadvantage must be considered. Patterns of technological superiority–inferiority are also significant *within* the Third World. Only a relatively small handful of countries (e.g. India, Mexico, Brazil) have the scientific capacity and resources necessary to launch a significant biotechnology programme. Further development of biotechnology and other information technologies is likely to reinforce existing tendencies towards differentiation within the Third World.

Corporate dominance in biotechnology

A second major dimension is that biotechnology, though in its infancy, is so extensively dominated by private sector R & D. Biotechnology represents the leading edge of a trend for private R & D in new technologies to be closely related to relatively "basic" research. Many basic research frontiers in molecular and cell biology have important commercial implications, and private-sector firms fund – and themselves often conduct – fundamental biotechnology research. The biotechnology industry was, in fact, "prematurely" commercialized, that is before the lines between basic and applied research were clear, and when major product revenues were generally more than a decade into the future. Premature commercialization was propelled initially by US venture capital which saw the opportunity for large capital gains, and later by the bitter competitive struggle among large multinational corporations (MNCs).

In most areas of biotechnology R & D worldwide, private firms are attempting to assert their dominance. In plant molecular biology the multinational and start-up companies are already directly dominant. Due to the lack of publicly funded research into the molecular biology of the higher plants in advanced countries, private firms have had to undertake much of this research in-house, or to fund directly research in universities, national laboratories and start-up companies. Private R & D expenditures in plant molecular biology in the US are estimated to be several times those of public agricultural research institutions. Monsanto and DuPont, two major US-based MNCs, together spend more on life-sciences research (in agriculture plus other product lines) annually than the total base budgets of the thirteen IARCs.

However, even in areas where public-sector research is more prominent, such as nitrogen fixation, photosynthetic efficiency and some aspects of animal agriculture and food biotechnology, the new milieu of international technological competition gives a quasi-private character to the results of this research. Universities are under increased pressure – both political and financial – to transfer research results more quickly and effectively to private industry, to patent new discoveries, and to license them to private firms to generate royalty income. Further, the majority of senior molecular biologists in universities are linked to private industry through exclusive consultancy relationships, equity ownership, membership of scientific advisory boards, or research grants. These factors serve to ensure that commercially relevant research results are patented (or are subject to trade secrecy) and preferentially available to private firms.

The significance for the Third World of private-sector dominance in biotechnology goes beyond the likelihood that research and production processes, genes, crop varieties and compounds will be patented or protected as trade secrets. The transformation of universities and other public research institutes in the advanced countries, engendered by the new era of international technological competition, suggests that Third World countries will face increased problems in securing technical linkages and co-operation with them.

The predominance of First World-orientated research priorities in biotechnology R & D

To a far greater degree than previous industrial technologies such as plastics and electronics, biotechnology is *potentially neutral* with regard to alternative social or development goals and the livelihoods of various classes of producers and consumers.[2] For example, scale economies in biotechnology are far less than in automobile or steel production. Likewise, biotechnology can be devoted to improving the production of luxury goods (e.g. "ice-minus" frost protection of high-value horticultural crops, expensive therapeutic drugs), wage goods (e.g. drought-tolerant cereal grain varieties, inexpensive vaccines for endemic infectious diseases) or subsistence crops (e.g. new cassava varieties). The tools of biotechnology potentially enable researchers to benefit either capital over labour or labour over capital, to benefit peasant farmers over their plantation counterparts or vice versa. Thus, the most crucial factor in shaping the socio-economic consequences of biotechnology innovation lies in the *structure through which research priorities are determined*.

However, current research priorities tend to be of limited applicability for the majority of Third World residents, if not adverse to their interests. The vast bulk of biotechnology R & D now occurs in the OECD countries, where most expenditures are directly accounted for by private-sector firms, with much public-sector R & D undertaken for the indirect benefit of private firms. Not surprisingly, the vast bulk of R & D is devoted to products intended primarily for the far larger markets and more affluent consumers in the developed world. While some biotechnology products currently under development in OECD countries will have potential markets and may confer benefits in Third World countries, this is more by accident than design.

Most proponents of biotechnology for development place particular stress on its ability to improve Third World health care and food production. While this is correct in principle, the determination of research priorities in the world biotechnology industry is such that much of the potential will go unrealized. Take pharmaceuticals and health care, for example, where the most pressing needs in the Third World are for basic public health services and inexpensive vaccines and other medicines. Multinational and start-up biotechnology companies, however, devote virtually all their attention to developing products such as high-value prescription drugs (e.g. tissue plasminogen activator), diagnostic kits and so on, aimed at large markets in the OECD countries where they will receive heavy subsidies through national health care programmes.

There are similar limitations on the development of agricultural biotechnologies in the North. Most of the first generation of R & D in agricultural biotechnology has been devoted to the production problems of First World farmers, those that occur in capital- and energy-intensive monocultural cropping systems and in large, capital-intensive confinement livestock production systems. These technical problems – for example, rationalizing chemical weed control, reducing the share of fat in swine carcasses – are generally of limited importance to Third World agriculture. First World agriculture is typically based far more on purchased off-farm inputs such as fertilizers, pesticides and crop varieties than Third-World – especially peasant – agriculture. Biotechnology R & D for agriculture is generally aimed at reinforcing the dependence of farmers on purchased inputs.

Because some pest and disease resistances can be introduced to crop plants as single-gene traits, some private firms devote part of their biotechnology R & D to non-chemical means of plant protection. This modest emphasis, however, is due to the simplicity and speed of introducing such single-gene traits into varieties, and to the tremendous pressure on private firms to get biotechnology products on the market. In addition, there are vast agronomic differences between the temperate agro-ecological zones of the developed countries and the tropical or sub-tropical agro-ecological zones of Third World countries, further limiting the relevance of agricultural biotechnologies produced in the North to the countries of the South.

Even where First World research addresses Third World applications, the concerns of market size and dependence on purchased off-farm inputs come into play. The increasingly competitive agricultural input industries of the North, with their trend towards market saturation, are giving attention to exports to Third World countries as the most promising frontier for market expansion.[3] As agricultural input sales depend upon a costly and extensive marketing network (in addition to costly R & D), potential sales revenues in the Third World must be sufficiently large and reliable to justify R & D and marketing investments. These requirements are best met by relatively large Third World countries with large numbers of highly commercialized producers of major Green Revolution cereal grains (wheat, rice, maize and sorghum). Biotechnology-derived crop varieties developed and sold will generally be hybrids, for both social and biological reasons.[4] Since most Third World countries have poorly developed plant breeder's rights and patent systems, hybrid varieties (which are reproductively unstable and do not breed true from generation to generation) represent a "biological patent". Hybrid varieties do not permit farmers to save seeds from the previous season's crop, compelling them to buy seed each growing season.

The hybrid varieties of maize and sorghum being sold in Third World countries such as Mexico, the Philippines, Colombia, Argentina and Brazil, and the wheat, rice and soya bean varieties that will probably soon be marketed in the Third World, will increase cereal grain output and reduce food prices. These benefits are likely to be localized, however, perhaps more than those of the Green Revolution on the heels of which the "Biorevolution" will closely follow. Farmers will also become more dependent on purchased inputs. Biotechnology will telescope the impact that market penetration had during the Green Revolution, unless there is a significant public research thrust to make appropriate biotechnologies available to smaller farmers and to those in less favoured agro-ecological zones. Most importantly, biotechnologies developed and transferred through private channels will be of little or no relevance to small-scale producers of crops such as cassava, potatoes and grain legumes that are not of commercial interest to biotechnology firms. Whether these producers will benefit from biotechnology applied to Third World agriculture will depend on the degree to which public-sector research institutions undertake research geared to the distinctive technical needs of smallholder peasants.

Biotechnology and the decline of Third World exports

It is now widely recognized that the initial applications of biotechnology in the Third World will be industrial tissue culture and bioprocessing, which will displace indigenous sources of plant-derived chemicals and materials. In most instances the development of an industrial substitute for a plant-derived product will result in the production process being transferred from the Third World to one or more developed-country factory locations. Industrial "substitutionism"[5] will reduce Third World export revenues and employment in these labour-intensive export sectors; this is already happening for plant materials such as sweeteners, fragrances and spices. Multinational and small start-up biotechnology firms from a variety of advanced industrial countries – often with assistance from public research programmes – have pioneered these R & D efforts.

For example, private firms have already initiated cell-culture production of shikonin (a dye and pharmaceutical) and vanilla. High-fructose corn syrup (HFCS) and other sugar substitutes (most recently aspartame) have already pared world demand for sugar, a major Third World export crop. Recombinant DNA production of thaumatin (a sugar substitute potentially more revolutionary than its predecessors) is within one or two years of commercial application. New industrial substitutes for cotton fibres are also being explored. On the more distant horizon is the possibility that protein engineering techniques can be applied to conversion of low-price oils (e.g. olive, sunflower and palm oils) into cocoa butter, the principal ingredient for manufacturing chocolate; even further on the horizon is research on utilizing cell culture for the "biosynthesis" of cocoa butter in a factory.

Each case will bring a significant impact, most immediately a price decline for the commodity on world markets and decreased employment in the primary industry. As these substitute products increase their market penetration, the long-term outlook is for sharp falls in, if not a total disappearance of, export revenues from the crop concerned, along with high levels of poverty and unemployment in the zones where the crop was formerly cultivated.

These impacts will be highly variable, however, depending on the importance of such industrial raw materials as sources of export revenue, the size of the country's national product, and the social structure of the crop production process. For example, major world cocoa suppliers, particularly Ghana and Cameroon, and to a lesser extent Ecuador, earn much of their foreign exchange from this crop and will be most dramatically affected by a future decline in cocoa export revenues. Other major cocoa suppliers – Brazil, Nigeria and Malaysia – are larger, wealthier countries with more diversified export portfolios, and will be less affected by declining cocoa exports. About 57 per cent of world cocoa production is accounted for by African countries, which generally have the least flexibility to absorb major losses in export revenues. Among major world suppliers, Brazil and Malaysia have plantation-dominated production systems, while other major suppliers have small-scale, peasant-dominated systems. Declines (or rapid technical change) in cocoa production will probably have more adverse impacts on small producers than on plantation capital (though the impact on plantation labour forces will be very substantial).

Some observers (such as Goodman *et al.*) argue that the most important long-term impact of biotechnology on food systems could be to supplant farming and agriculture as the source of food commodities.[6] That is, the ultimate extension of biotechnology substitution could be increasingly to shift food production into the factory through expanded use of cell and tissue culture production and industrial bioprocessing techniques, thereby relegating agriculture to a source of feed-stocks for industrial bioprocessing. Such a prospective trend is guesswork at this point. None the less, it is important to recognize that although the lion's share of agricultural biotechnology R & D is currently focused on plant and animal improvement, industrial substitutes for agriculturally derived products will have a significant impact on many Third World countries. Moreover, for some (mainly African) Third World countries, "industrial substitutionism" will be the *initial* (and generally adverse) impact, and perhaps the only one for a decade or more. Biotechnology thus raises the possibility of a significant restructuring of the world food economy caused by the possible industrialization of food production, and the relegation of agriculture to production of biotechnology feedstocks.

Biotechnology product imports

The considerable potential long-term benefits of biotechnology to the Third World will come at a price, not least increased dependence on imports of biotechnology products, and of the capital goods and other producer goods and services necessary for R & D and production. For Third World countries with substantial research capacity in molecular and cell biology, biochemistry, immunology and related areas, investments in national biotechnology R & D programmes – including outlays for the imported capital goods, inputs and services required – will be clearly justifiable. National investments should enable these countries to set their own research priorities for the development of industries to meet national development goals, though of course this will not necessarily be the case. Most importantly over the long term, national biotechnology programme investments would reduce hard currency spending on biotechnology product imports. For the more numerous countries that lack such research capacity, however, imports of biotechnology products will be the major way, if not the only way, to secure access to these new technologies. Finally, a substantial group of countries, mainly those of the so-called Fourth World, but possibly some small, middle-income Third World countries with staggering debt loads and balance of payments problems, will be so strained in import capacity that most new biotechnologies will be unavailable to them.

MNC investments in biotechnology R & D, production and marketing

In biotechnology as in most other spheres, MNC investments in the Third World are a two-edged sword: these investments may contribute to overall economic expansion, make available needed consumer and producer goods, and help to transfer new technologies to the Third World; on the other hand, they may outcompete or weaken domestic capital, drain capital through repatriation of profits, increase external influence on economic policy, and contribute little to indigenous technical capacity. In most instances MNC investments involve some trade-off between growth and "dependence", which is shaped by the class and political structure and the world economic position of Third World countries, and by the nature of state policy towards foreign corporations.

Gereffi's useful comparative analysis of the pharmaceutical industry in fourteen countries in all three major Third World continents provides some tentative indications of growth – dependence trade-offs that might be played out with respect to the biotechnology industry.[7] The Third World as a whole occupies a contradictory position in the marketing strategies of MNC pharmaceutical firms. On the one hand, Third World drug sales, much like Third World seed and agricultural input sales, represent a significant frontier for MNC expansion. In the late 1970s, however, aggregate Third World pharmaceutical sales by MNCs accounted for only slightly over 11 per cent of the total, a relatively trivial amount.

Pharmaceutical MNC involvement in Third World economies has invariably generated major conflicts over drug prices, drug quality, and the extent of local production of active ingredients, and of technological transfer to the host country. Larger countries with larger markets have been in the best position to develop policies that effectively channel these MNC investments to achieving development goals, to regulate drug prices, to sustain locally owned pharmaceutical firms, and to establish state-owned firms to compete with MNCs or control them through joint ventures. Weaker countries of the South experience the most unfavourable growth – dependence trade-offs, as they have less bargaining power in two respects. Because they are less important markets, MNCs are less willing to negotiate arrangements more favourable for low-income countries; second, these countries have less powerful, less autonomous states – and accordingly less ability to bargain effectively, to support locally owned firms, or to finance the establishment of state-owned pharmaceutical enterprises.

Of the major production sectors affected by biotechnology, pharmaceutical MNCs currently have the largest presence in the Third World, followed closely by petroleum and food manufacturing firms, and then chemical and agricultural inputs firms. Direct MNC investments in Third World biotechnology production and marketing seem likely to follow these longstanding patterns (albeit with some increase in the role of agricultural-input MNCs over time), involving trade-offs similar to those discussed above.

The perverse timing of biotechnology applications in the Third World

Current development of the world biotechnology industry suggests the likelihood of perverse timing of biotechnology applications in most Third World contexts. Low-income countries seem likely to absorb the costs of biotechnology innovation initially, while the benefits will be slower in coming.

The initial impact of biotechnology on the Third World will be a decline in prices, export revenues, employment and income as a result of industrial biotechnology substitutions, as discussed above. The second stage of the impact, over the next ten to fifteen years, will probably be increased imports of biotechnology products made in the North, many developed primarily for First World applications and inappropriate for Third World development needs. These imports, which will yield major benefits but also entail costs such as demands on hard currency reserves and the weakening of domestic industry, will consist primarily of pharmaceuticals and agricultural inputs.

The final stage of biotechnology applications in the Third World will most likely follow contemporary patterns of conventional pharmaceutical production in the South. All things being equal, the twelve or fifteen largest, most affluent Third World countries will have their own state-owned or privately owned biotechnology firms producing both intermediate and finished products, and with their own R & D – in most instances as competitors or joint-venture partners of MNCs. The same countries will also have considerable public-sector (and possibly local private-sector) capacity to undertake biotechnology R & D for agricultural improvement. Several dozen more countries will have the capacity to manufacture and process several simple bulk drugs or chemicals, but with little or no R & D capacity. Their public agricultural researchers will be able to do routine biotechnology research, and to utilize biotechnology innovations developed elsewhere.

The remaining countries with no manufacturing facilities will depend on imported products in their finished forms. Their research institutions will, at best, have the capacity to make use of plant and animal materials that contain novel genes.

At this concluding stage of biotechnology application in the Third World, the extent of benefits accruing to the South will tend to be very unequally distributed, roughly according to the existing patterns of wealth and accumulation. This assessment has two qualifications. First, the sharply unequal distribution of biotechnology benefits within the Third World *is* amenable to public policy interventions, by countries of the South themselves and by multilateral development institutions. Second, the next twenty or thirty years are likely to see patterns of upward and downward mobility among Third World nations, much like those that have occurred since the Korean War. The rise of the Asian Newly Industrializing Countries (NICs) was totally unanticipated two or three decades ago, as was the downward mobility experienced by countries such as Argentina. The likelihood of mobility in the world economic and social structure, however, should not deflect attention from a fundamental reality: that the socio-economic forces at work today portend a temporally perverse and spatially unequal distribution of the costs and benefits of biotechnology in the South, unless there are policy interventions capable of modifying them.

The formation of world markets in biotechnology feedstocks

Most analysts of emerging biotechnology industries focus almost entirely on what they produce, and very little on the nature of the inputs required. My concern here is with the often-neglected, longer-term resource requirements for the world biotechnology industry. Plant-derived feedstocks are almost certain to become a central input, assuming that biotechnology lives up to its promise to provide substitutes for fossil hydrocarbon feedstocks in producing chemicals and fuels.

A full-blown global biotechnology industry twenty or thirty years hence, in which the bulk of pharmaceutical production and significant shares of chemical and energy production are based on plant-derived materials, would represent a dramatic transition in the use of primary biological materials. Since most production facilities will be in the North, the largest share of feedstocks (grains, cellulosic materials and so on) are likely be produced in the North as well, particularly if the advanced countries continue to face agricultural overcapacity and overproduction problems.

While low-income countries are unlikely to become the major suppliers of biomass feedstocks, the increased importance of these materials for industry could have significant implications for the South. Third World economies – especially their rural sectors – are generally heavily based on primary production from ecosystems. There is growing concern that the use of farmland, forests and other ecosystem resources in much of the South is already unsustainable. The diversion of even modest amounts of grains, forest materials and other organic matter to industrial production would place additional pressures on the primary resource base.

Further, it is possible that world agricultural product markets may become transformed, in part, into markets in *feedstock materials*, with Japan, the Asian NICs and some European countries being the most likely destinations. These foreign demands on Third World ecosystem productivity would add to the pressures of indigenous demand. It is even possible that grains will, in effect, be bid away from Third World consumers on world markets because of their value as industrial feedstocks in a mature biotechnology industry. This aspect of Third World dependence in biotechnology, though clearly only long-term and *hypothetical* at present, has none the less been ignored almost totally by social and ecological researchers.

Conclusion: biotechnology and the future of Third World development

Biotechnology represents a challenge and an opportunity for countries of the South. On the hand, the "biorevolution" threatens to exacerbate North-South technical disparities, reinforce patterns of Third World dependency and inequality, and create even more pronounced socio-economic differentiation among low-income countries, as many critics have warned.[8] On the other hand, biotechnologies in agriculture, non-farm industry, health care and so on could make major contributions to improving living standards and enhancing accumulation in the Third World. But they will be able to do so neither automatically nor painlessly.

Perhaps the most crucial aspect of biotechnology for development by comparison with previous industrial technologies is that it is relatively inexpensive and thus potentially accessible to most developing nations.[9] By contrast with a "turnkey" steel, oil refining, auto assembly or chemical facility costing upwards of several hundred million dollars, a state-of-the art biotechnology research facility and a pilot bioprocessing production facility could be established with a capital investment of $50 million. Relatively inexpensive biotechnology facilities can also provide R & D for many sectors – agriculture, food manufacturing, pharmaceuticals, energy, forestry, mining, chemicals – while a conventional industrial plant is limited to one sector. Especially for the poorest Third World countries, $50m is a major government commitment; even more would be needed for scale-up and commercialization of biotechnology products. But given the "multivalency" of R & D facilities, these investment requirements are modest. Unfortunately, the political economy of international biotechnology will tend to limit the development gains possible. Third World countries will have to do more on their own – either individually or as members of Third World consortia – than they have in the past, in order to establish biotechnology R & D programmes that can contribute to achieving equitable development.

Notes

1. "Is biotechnology a blessing for the Third World?", *Monthly Review*, no. 34 (1983), pp. 10–19. See also Martin Kenney, *Biotechnology* (New Haven: Yale University Press, 1986).
2. This notion of the *potential* socio-economic neutrality of biotechnology does *not* imply that this technology will ultimately be neutral in its impacts. Biotechnology research priorities and products, much like those from most other major technologies, will be about as socially just in their impacts as the societies in which they are deployed.
3. The attention paid by MNCs to developing agricultural biotechnology products – especially new crop varieties – for export to Third World countries has declined over the past three or so years (and this research emphasis is now less than was predicted a few years ago by, for example, Kenney, *op. cit.,* and myself). In substantial measure this decline reflects a recognition of the technical difficulties of developing viable crop varieties for sale in the Third World, the likelihood of competition (with domestic firms and national agricultural research institutes, as well as with other MNCs) over limited markets, and the expense of creating marketing infrastructures in several low-income countries.
4. See Jean-Pierre Berlan and R. C. Lewontin, "The political economy of hybrid corn", *Monthly Review*, no. 38 (July–August 1986), pp. 35–47. For a an important extended treatment, see Jack Kloppenburg, Jr, *First the Seed* (New York: Cambridge University Press, 1988).
5. The notion of substitutionism – the application of technology to displace "natural" processes for producing food and fibre with industrial processes – and arguments concerning how biotechnology may accelerate substitutionism have been elaborated by David Goodman, Bernardo Sorj and John Wilkinson, *From Farming to Biotechnology* (Oxford: Basil Blackwell, 1987).
6. ibid.
7. Gereffi, Gary, *The Pharmaceutical Industry and Dependency in the Third World* (Princeton, NJ: Princeton University Press, 1983).
8. Among the more important popularized critiques of the political economy of biotechnology and Third World agricultural

development are Jack Doyle, *Altered Harvest* (New York: Viking, 1985) and Henk Hobbelink, *New Hope or False Promise? Biotechnology and Third World Agriculture* (Brussels: International Coalition for Development Action, 1987). Another useful resource is Research and Information System for the Non-aligned and Other Developing Countries (RIS), *Biotechnology Revolution and the Third World* (New Delhi: RIS, 1988), which contains an important set of scholarly but accessible papers.

9. See Frederick H. Buttel and Martin Kenney, "Biotechnology and international development: prospects for overcoming dependence in the information age", in D. F. Hadwiger and W. P. Browne (eds), *Public Policy and Agricultural Technology* (New York: St Martin's Press, 1987), pp. 109–21; and Martin Kenney and Frederick H. Buttel, "Biotechnology: prospects and limitations for third world development", *Development and Change*, no. 16 (1985), pp. 61–91.

15 Urban Consumptionism as a Route to Rural Renewal

Robin Jenkins

If the world's agricultural land is divided by our human population we each have about one hectare, of which one-third is arable and two-thirds is pasture; that is, a patch of grass 80 metres square plus a patch of earth 60 metres square. It is not much but it is more than adequate to grow a healthy diet – with love and care it is enough to feed ten of us; but it is not enough to grow the unhealthy diet of just one person in an industrialized nation. This is because our unhealthy industrialized diets overdose on fat and protein, both of which require more square metres of land per calorie of food than do basic carbohydrates.

The demand for land is even more exaggerated if most of the protein is required from animal sources. This is particularly the case with beef, which has a poorer conversion ratio from vegetable protein to animal protein than pigs or sheep or poultry. Anyone who eats a half-pound burger every day is using a good 8 hectares of land – and not just pasture because typically, burger beef has to be fed concentrates grown on arable land, in addition to grass.

As a species we humans are increasingly divided between those who die prematurely from overconsumption of foods that make us ill, using far more than our fair share of land in the process, and those who die prematurely from straightforward lack of important nutrients, being too poor to lay claim to the produce from that one hectare that is their rightful share.

There are grounds for hope within this predicament because it is clear that everyone could gain. If the richer half of the world were to improve its quality of life by eating more healthily, then less land in the Third World would be dedicated to the production of export beef, export peanuts or export oil and oilseed cake, either for direct consumption by the rich world or to fatten their already fat animals. This does not necessarily mean that the spare land will then be used to grow food for the poor, but it does at least halt the flow of fat and protein from nations where there is a net deficit to nations where there is a harmful and unhealthy surplus. Until this flow of food is stopped there seems to be little likelihood that the totally false belief in economic development via increasing food exports will ever be seen as the hopeless mirage that it really is. Cuba, Nicaragua, Botswana, Senegal, Brazil. . .the list is endless and crosses the ideological spectrum. . .all strive after this mirage. There is little hope that any Third World regime can or will stop the flow of fat and protein, so we have to look back to the metropolitan centres for possible solutions.

The argument that follows uses a very simple logic, and if action can be taken in Britain it can be taken anywhere. Britain's main claim to fame concerning food is its advanced development of the insane logic of industrialized food production under a regime publicly committed to Friedmanite economics. This results in deep and fundamental contradictions which it is our duty to expose. The global dislocation between food production and food needs has been the consequence of imperialist relations and military strategies (both attack and defence) more than of factors such as climate or soil. This dislocation takes on a particular and advanced (as in advanced cancer) form in Britain due to two factors related to its imperialist history.

One, Britain was the first country in the world where the industrial bourgeoisie won its political ascendancy over the old landowning classes. This victory is usually associated with the repeal of the Corn Laws which was followed by the import of cheap grain and a slump in British production. By the end of the nineteenth century, agriculture had become an addendum to the British economy while production of food for Britain had become big business in settler territories like the USA, Canada and Australia, and had resulted in indentured labour plantations throughout much of the Third World.

Two, Britain was also the first country in the world to force its people off the land and into the cities, with the result that most indigenous families now have to go back at least five generations to discover any contact with the land. Although some 98 per cent of the people own no agricultural land, they do not think of themselves as landless, and most have little or no knowledge of growing food. In fact, recent immigrants to Britain's inner cities from Bangladesh, Guyana or Jamaica often have more recent roots in the land and more understanding of agriculture and its politics.

This combination of early and overwhelming urbanization accompanied by cheap food imported from all over the world made it possible for the ascendant bourgeoisie to depoliticize food, and largely remove it as an issue from national politics by the outbreak of World War I. Ironically, removing food from the political agenda was actually compounded when food supplies were threatened during the war, because the government set up a Ministry of Food to plan production according to needs. Afterwards British agriculture was allowed to sink back into obscurity until World War II when a Ministry of Food was again set up to plan food production and ration consumption according to need, a system that continued through to 1954.

After 1954, however, there was no return to the free trade policies that had held – except during wartime – since the repeal of the Corn Laws in 1846. In fact, British agriculture has been maintained on a war footing ever since World War II despite the construction of nuclear weapons and the strategy of threatening mutual destruction known as "deterrence", which has removed conventional warfare from the European agenda. The landed gentry (mostly by now also industrial bourgeoisie) regained all that had been lost with the repeal of the Corn Laws a century previously and grew fat and prosperous on import controls, subsidized inputs and guaranteed prices for all outputs, irrespective of food needs or even market demand.

Such absolute guarantees inevitably led to a massive industrialization of agriculture in a headlong frenzy for constantly increased production. More and more fertilizer was thrown on the land to increase yields. As fertilizers are subject to decreasing marginal returns, so exponential increases in inputs were required to produce uniform increases in the harvest. If British agriculture were now to be exposed to free international competition, as Thatcherite/Friedmanite economics dictates it must, just about every farm in Britain would have to shut down within weeks.

The British diet: nutritional squalor amidst plenty

While British food production has now been on a war footing for fifty years, ever since 1954 food consumption has been left to the vagaries of a free market without standards or controls. This has resulted in nutritional squalor amidst plenty. The litany of statistics describing this bizarre and immoral situation is a long one.

British food production is:

(1) the most subsidized in the world
(2) the most inefficient in the world in terms of energy inputs and outputs
(3) the most dependent on fossil fuels in the world
(4) the highest user of fertilizers in the world
(5) the most mechanized in the world
(6) the most inefficient in the world in terms of cost inputs and value outputs
(7) the most packaged and processed in the world.

British food is:

(8) the most adulterated in the industrialized world
(9) nearly the most fatty in the industrialized world.

The British suffer:

(10) the highest incidence of diet-related disease in the world
(11) the most expensive food in the world, including that element of taxation required to fund farming subsidies and price support
(12) the worst deterioration in their diet of any country since World War II.

Let us now go through this bizarre list point by point.

(1) Subsidy

Subsidies and public costs of food production in Britain take a number of forms. The Ministry of Agriculture costs some £800 million per year, a twelvefold increase in real terms over the fifty years from 1938, when it was £60 million at today's prices.

UK expenditure on price support for agricultural products now averages almost £300,000 per farmer per year. Over the past thirty years the total has more or less doubled from £1,696 million in 1955 to £3,150 million in 1983 (stated in 1983 prices). Over the whole of this period of price support, starting in 1947, it is calculated that up to 1987 the total paid out in price support has amounted to at least £75,000 million, almost all of which can be accounted for as the artificial increase in the value of agricultural land over the same period. In other words, capital has flowed into the purchase of farmland on the assumption of permanent and increasing price support. This has little to do with any free market mechanism.[1]

The Institute for Fiscal Studies calculates the current cost of protectionism in the form of import controls, tariffs and levies as £2,000 million in government expenditure and £3,000 million extra on the cost of food. This represents some 12 per cent of total expenditure on food or £5 per week for an average family. Another striking way of evaluating this £3,000 million is to compare it with the value of home farm output at around £6,000 million. The extraordinary fact is that these subsidies, supports and protections increased some 14 per cent in real terms in just the first two years of Thatcher's avowedly Friedmanite administration.

(2) *Energy inefficiency*

The energy input for food production in Britain averages about one ton of oil per hectare of agricultural land. (A hectare is 100 metres by 100 metres.) Until very recently, fertilizer consumption was increasing at the rate of 5.3 per cent per year, so it doubles every fifteen years. Food production and distribution currently consume some 26 per cent of all energy consumed in Britain and the continuation of industrialized agribusiness must lead to a steady increase per hectare, per person, per year. In terms of energy inputs and outputs the ratio has deteriorated from around 1:4 (input:output) in 1945 down to a little more than 1:2 in 1985. In other words, British agriculture has declined to half its efficiency during World War II. The explanation of this process is simple: fossil fuels at non-replacement costs have been substituted for labour which is comparatively more expensive. Table 15.1 shows changes in the efficiency of maize production in the United States between 1945 and 1970, indicating the processes at work. The figures are a recalculation from a study reported in Pimentel,[2] so as to provide a direct comparison of the energy inputs required in 1945 and 1970 to produce 1 million calories of maize. The cost of the increased labour productivity seems quite incredible.

Table 15.1: Energy inputs to produce one million calories of maize (calories)

	1945	*1970*	*efficiency % change*
Labour	3,548	600	+508
Machinery	52,524	51,445	+2
Fuel	158,448	97,624	+62
Fertilizers			
Nitrogen	17,158	115,238	−672
Phosphorus	3,093	5,769	−83
Potassium	1,517	8,329	−549
Seeds	9,921	7,716	
Irrigation	5,544	4,165	+33
Insecticides	0	1,347	
Herbicides	0	1,347	
Drying	2,918	14,699	−80
Electricity	9,338	37,972	−75
Transportation	5,836	8,574	
Total energy input	269,945	354,825	
Total energy output (Maize)	1,000,000	1,000,000	
Energy ratio (in/out)	1:3.7	1:2.8	31% decline in efficiency

Source: Recalculation from figures given in D. Pimental and C. Pimental, *Food, Energy and Society*, 1979.

(3) *Fossil fuel dependency*
This can be explained in terms of Table 15.1. Britain uses less labour per calorie of harvest than any other country, has the most mechanized agriculture, comes fourth in the use of fertilizers per hectare, and is among the top five users of insecticides, herbicides, etc. Bearing in mind that energy inputs are generally in line with those given for maize, then the use of nitrogenous fertilizers is the biggest input, followed by fuel, and then machinery.

(4) *Fertilizer use*
Japan, West Germany, the Netherlands and Belgium all use more fertilizers per hectare than Britain but this is because Britain has a higher percentage of pasture than these countries. If fertilizer use is calculated separately for pasture and arable land (and it has to be said that the calculation requires some assumptions), then Britain comes out as the highest user of fertilizers on pasture and third highest on arable land.[3]

(5) *Level of mechanization*
Britain does not come top in the tractor stakes but if combine harvesters and milking machines are added into the equation, then the use of machinery per hectare is the highest in the world, with almost two tractors per farm labourer.[4]

(6) *Cost efficiency*
Many studies have now been made of the comparative efficiency of different farming methods, including international comparisons. Perhaps the most reliable recent figures are contained in the *Global 2000 Report to the President*,[5] though this focuses in the main on energy rather than economic values. It is clear that the whole British farming sector only stays in business because of massive protectionism over and above that offered by

Table 15.2: Consumption of healthy and harmful foods (1945–73)

Foodstuff	*No limits*	*Harmful in excess*	*Consumption 1945–73*	
Milk		X	down	1.6%
Cheese		X	up	47.5%
Butter		X	up	14.9%
Margarine		X	up	23.1%
Sugar		X	up	35.0%
Meat		X	up	24.3%
Fish	X		down	30.2%
Potatoes	X		down	9.6%
Bread	X		down	39.8%
Vegetables	X		up	12.0%
Fruit	X		up	38.8%
Eggs		X	up	20.8%

Note: If there are no limits to consumption the foodstuff scores an "X"; likewise if it is harmful in excess. The column to the right shows what has actually happened in the postwar period up to 1973, since which time consumption of sugar, butter, cheese and eggs has continued to increase.

the EEC; although there are no directly comparable international studies which successfully compare like with like, there are no other countries so dependent on both price support and import controls.[6]

(7) *Packaging and processing*

The British eat a higher percentage of processed food, 72 per cent, than any other nation, including the USA. In other words, more food comes out of cans, packets and freezers than elsewhere. Furthermore, Britain has the worst record of all the industrialized nations for recycling such packaging, according to a recent market research report.[7]

(8) *Adulteration*

Adulteration is a dirty word but it is used advisedly here to mean the addition to food of substances that are known to be harmful, which consumers have not specifically asked for and which are not necessary. These include sugar, which is added to an alarming array of processed foods such that at least half of British sugar consumption is indirect and unnoticed by the consumer – to the extent of an average 50 lbs per person per year.[8]

(9) *Proportion of fat*

Only four nations exceed Britain's per capita consumption of fat but the comparative statistics do not differentiate between the more and less harmful fats – saturated, polyunsaturated or mono-unsaturated. There is, however, independent evidence collected by the Nuffield Institute of Zoology which suggests that the British eat a higher percentage of the more harmful fats than any other nation.

(10) *Incidence of diet-related disease*

There is of course room for debate on both the incidence of diet-related disease and the degree to which such diseases are caused by diet and/or other factors but the most reliable analysis available of the British statistics is that carried out by William Laing, ex-Deputy Director of the Office of Health Economics in the old Department of Health and Social Security.[9] Per capita comparisons of such diet-related diseases as dental decay, anorexia, intestinal cancer, diabetes, diverticular disease, heart disease, gall bladder disease, high blood pressure, obesity, strokes, mineral deficiencies and vitamin deficiencies consistently place Britain at or very near the worst extreme on all counts.

(11) Cost to the consumer

The real costs of food in Britain are high, due to inherently expensive and energy-inefficient means of production, compounded by price guarantees and the protectionism that alone allows such an internationally uncompetitive industry to stay afloat. It is, however, necessary to distinguish between international comparisons of food costs at official exchange rates (including hidden costs paid in taxes), where Britain scores very badly, and international comparisons of the percentage of disposable income spent on food, where many nations are of course paying out a much higher percentage than the British.[10]

(12) Deterioration in diet

It is normal to divide foodstuffs into twelve categories, of which five are beneficial in unlimited quantities while the other seven are all harmful when eaten to excess. They are as shown in Table 15.2. There is no other known example of any country where the diet has deteriorated so sharply over the same period. Partly this is because the British diet under war rationing was both adequate and healthily balanced, albeit boring and unvaried.[11]

The Greater London Council: intervening for a better diet

Food capital in Britain has conglomerated rapidly into a corporate structure of vertically integrated investments – from seed breeding and chemical inputs right the way through to the processed, packaged, canned or frozen edibles. Throughout, under both Tory and Labour governments, there has been a sweetheart relationship between monopoly food capital and the Ministry of Agriculture. Neither consumer nor worker interests have ever been represented on any of the supposedly regulatory bodies, all of whose proceedings are shrouded from public view by the protection of the Official Secrets Act. It is still treacherous under the law to reveal the deliberations of such bodies. Until 1984 the main source of public information on food and nutrition was provided by the British Nutrition Foundation, an organization 98 per cent funded by the food industry which purveyed only those facts compatible with making a good profit. This scandalous combination of vested interests and preventable disease was not a public issue because the historical depoliticization of food was accompanied by apparently cheap prices in the shops, with the real costs obscured by government intervention.

In 1983, the Greater London Council (GLC), which was the elected city government of some seven million Londoners, decided to intervene.[12] It adopted a food policy which aimed to make considerable use of its purchasing powers (including food for over one million school meals per week), its training powers with the Greater London Training Board, its banking and investment powers with the Greater London Enterprise Board and its planning powers through its Planning Committee. The GLC was a radical Labour administration which strongly believed in augmenting its formal powers with popular involvement and public campaigning as the best guarantee of actually delivering on its stated policies. To this end it set up the London Food Commission (LFC) as a publicly accountable counterpart to the British Nutrition Foundation. The LFC was set up as an alliance between food workers and food consumers, and its company directors were drawn in equal part from trade union and consumer organizations.

The GLC was abolished by the government in the spring of 1986. It had been too successful at showing there was an alternative to Thatcherism; the lady could not tolerate such opposition so London no longer has a city government.

Despite threats of legal action it proved possible to forward-fund the London Food Commission through into 1990, so although the formal part of the food policy fell when the GLC was abolished, the campaigning part was able to continue, much to the annoyance of the food industry which coined a new term of political abuse, "Food Leninist".

After some five years of sustained research, debate, publications and controversy, the politics of food has become a major national issue in Britain, resulting recently in the resignation of a Minister for the Distribution of Preventable Disease (Health) and the setting up of a Cabinet Review chaired by Thatcher herself. Neither the government nor food capital has managed to extricate itself from a series of scandals on food adulteration, food poisoning, radioactive sheep (post Chernobyl), food irradiation, cook-chill guidelines, the EEC agricultural surplus, nitrate pollution of drinking water and the worsening statistics on the incidence of diet-related diseases.

While millions of pounds were being knocked off the share value of the food producers and processors, the industry and the government got locked into a public exchange of mutual recriminations. One sector of big food capital has profited from these developments, which might not themselves have been possible without a tacit alliance between the so-called "Food Leninists" and the vast purchasing power of the supermarket chains. It was a stroke of luck that in Britain, food retailing capital has remained quite separate from food production and processing capital. This meant that as soon as food issues got into the public domain the supermarkets were free to adjust their food purchasing and shelving accordingly, at little or no cost to themselves. In fact the intense competition between the dominant five supermarket chains served to accelerate the process, with each scrambling to be ahead on any issue, be it policy on irradiated food, stricter standards on pesticide residues, banning of suspect additives or stocking organic produce. Of course, once it became clear that the supermarket chains could and would respond, the pressure increased. It is significant that consumer demands have become more subtle and far-reaching, and now include demands for full information on food contents, and production methods, especially where animals are involved. Meanwhile, the food industry trade unions are demanding a say in the direction of new investment in production and processing on the grounds that jobs in a factory producing junk food are no longer safe, and companies need conversion plans to healthier products with a more assured long-term market.

Just about everyone in Britain now understands the relationship between diet and health, though many are slow to change their eating habits accordingly. Fewer people understand the relationship between diet and land, and indeed some city children are disgusted when they are first told in school that much of their food comes out of the ground. Even fewer people understand the relationship between diet, land, work and exploitation, or have any idea of the oppression involved in producing a pineapple or a bunch of bananas for their dinner table.

The grounds for making such links at the dinner table or at the supermarket are more fertile than they were. It is anyone's guess whether the changes that are coming about in British food consumption arise from selfish self-interest and the fear of diet-related disease or from social or ecological concerns about producers or production methods; suffice to say that there are at least two good reasons to stop eating most foods that are decidedly unhealthy. First, if you eat too many beefburgers you will overdose on fat, salt and protein while denying yourself fibre, minerals and vitamins. Second, if you eat too many beefburgers you will be using more than your fair share of the world's agricultural land, and more specifically you will be adding to the destruction of Amazonian rainforest by increasing ranching in Rondonia. But if you eat fewer beefburgers you can improve your own diet and health while, at the very least, liberating meat for those who do not get any, or releasing land hitherto used to produce concentrated cattle foods for the production of food for people, or saving virgin land from ecological degradation.

Britain is not the most politically fertile territory for making such links and achieving such changes but they are nevertheless happening. If they can happen in Britain, they can happen anywhere. That means that dietary improvements in the rich countries can become the motor for agricultural improvements in the poor countries which are currently tied into the export of protein in particular and calories in general and which actually suffer a net deficit.

Notes

1. For a refreshing critique of such policies it is difficult to beat the book by a Tory MP, Richard Body, *Farming in the Clouds* (London: Temple Smith, 1984). (Unfortunately his solution is laissez-faire capitalism.)
2. Pimentel, D., and M. Pimentel, *Food, Energy and Society* (London: Edward Arnold, 1979) provides a good introduction to the energy efficiency of food production. The figures in Table 15.1 are recalculated from data presented in D. Pimentel *et al.*, "Food production and the energy crisis", *Science*, no. 182, pp. 443–9, 1973.
3. The statistics used for this calculation come from the UN Food and Agriculture Organization, *Annual Production Yearbook* (Rome, 1988).
4. Ibid.
5. US Department of State, *Global 2000 Report to the President* (Washington, DC, 1982).
6. The trilogy by the sacked Chairman of the Parliamentary Select Committee on Agriculture provides a wealth of damning evidence on the inefficiency of British agriculture as a sector and its total dependence on the two crutches of price support and protectionism. See R. Body, *Agriculture: the Triumph and the Shame* (Aldershot: Temple Smith, 1982); op.cit., 1984; *British Agriculture: from Red to Green* (Aldershot: Temple Smith, 1987). See also note 1 above.
7. See *Environment Digest*, no. 27, p. 12.

8. For an analysis of the chemicals permitted in Britain but banned elsewhere there is no better source than *Food Adulteration – How to Beat It* (London: London Food Commission, 1988).
9. This work was actually commissioned by the *Sunday Times* and published in the issue dated 24 July 1983.
10. Body, op.cit., 1982, 1984, 1987.
11. The *Annual Report* of the National Food Survey, published each year by the DHSS is the source for most consumption statistics.
12. The main statement of GLC food policy is contained in a report to the Industry and Employment Committee in March 1984 entitled *Food for a Great City*. Copies may be obtained from the London Food Commission, 188 Old Street, London EC1.

16 Further Resources on Food

Charlotte Martin

Some Useful Books

(Nearly all of these are available in paperback.)

Agarwal, Bina, *Cold Hearths and Barren Slopes. The Woodfuel Crisis in the Third World* (London: Zed Books, 1986).
Agarwal, Bina (ed.), *Structures of Patriarchy. The State, the Community and the Household* (London: Zed Books, 1989).
Blakie, Piers, *The Political Economy of Soil Erosion in Developing Countries* (London: Longman, 1985).
Burbach, Roger, and Patricia Flynn, *Agribusiness in the Americas* (New York: Monthly Review Press, 1980).
Castro, Josue de, *The Geopolitics of Hunger* (New York: Monthly Review Press, 1977).
Christodoulou, Demetrious, *The Unpromised Land. Agrarian Reform and Conflict Worldwide* (London: Zed Books, 1989).
Crow, Ben, and Alan Thomas, *Third World Atlas* (Milton Keynes: Open University Press, 1983).
Crow, Ben, Mary Thorpe *et al.*, *Survival and Change in the Third World* (Cambridge: Polity, 1988).
Dankelman, Irene, and Joan Davidson, *Women and Environment in the Third World* (London: Earthscan Publications, 1988).
Dinham, Barbara, and Colin Himes, *Agribusiness in Africa* (London: Earth Resources Research, 1983).

George, Susan, *How the Other Half Dies. The Real Reasons for World Hunger* (Harmondsworth: Penguin, 1977).

Hartmann, Betsy, and James Boyce, *A Quiet Violence. View from a Bangladesh Village* (London: Zed Books, 1983).

Jackson, Tony, with Deborah Eade, *Against the Grain* (Oxford: Oxfam, 1982).

Johnson, Hazel, and Henry Bernstein, *Third World Lives of Struggle* (London: Heinemann Educational, 1982).

Juma, Calestous, *The Gene Hunters. Biotechnology and the Scramble for Seeds* (London: Zed Books, 1989).

Mackintosh, Maureen, *Gender, Class and Rural Transition: Agribusiness and the Food Crisis in Senegal* (London: Zed Books, 1989).

Moore Lappé, Frances, and Joseph Collins, *World Hunger: Twelve Myths* (London: Earthscan Publications, 1988).

Moore Lappé, Frances, and Joseph Collins with Cary Fowler, *Food First. Beyond the Myth of Scarcity* (New York: Ballantine Books, 2nd edn, 1979).

Morgan, Dan, *Merchants of Grain* (London: Weidenfeld and Nicolson, 1979).

Pacey, Arnold, and Philip Payne, *Agricultural Development and Nutrition* (London: Hutchinson, 1985).

Raikes, Philip, *Modernising Hunger. Famine, Food Surplus and Farm Policy in the EEC and Africa* (London: James Currey/CIIR, 1988).

Redclift, Michael, *Sustainable Development. Exploring the Contradictions* (London: Longman, 1987).

Richards, Paul, *Indigenous Agricultural Evolution* (London: Hutchinson, 1985).

Sen, A. K., *Poverty and Famines* (Oxford: Clarendon Press, 1981).

Sen, Gita, and Caren Grown, *Development, Crises and Alternative Visions: Third World Women's Perspectives* (London: Earthscan Publications, 1988).

Shiva, Vandana, *Staying Alive. Women, Ecology and Development in India* (London: Zed Books, 1988).

Timberlake, Lloyd, *Africa in Crisis* (London: Earthscan Publications, 2nd edn, 1988).
Wisner, Ben, *Power and Need in Africa. Basic Human Needs and Development Policies* (London: Earthscan Publications, 1988).

Journals and Periodicals

Journals

Food Policy
World Development
Journal of Peasant Studies
Journal of Contemporary Asia
Review of African Political Economy

Periodicals

Ecologist,
Worthyvale Manor Farm,
Camelford,
Cornwall PL32 9TT,
UK

Food Matters Worldwide,
Development and Environment Centre,
38–40 Exchange Street,
Norwich NR2 1AV,
UK

New Internationalist,
42 Hythe Bridge Street,
Oxford, OX1 2EP,
UK

Useful Organizations

International

UN Food and Agricultural Organization,
Via delle Terme di Caracalla,
00100 Rome,
Italy

Food First,
Institute for Food and Development Policy,
1885 Mission Street,
San Francisco, CA, 941023,
USA

Hunger Exchange,
Brown University,
Campus Box 1831,
Providence, RI, 02912,
USA

Institute for Development Anthropology,
99 Collier Street,
PO Box 818,
Birmingham, NY, 13902,
USA

International Food Policy Research Institute,
1176 Massachusetts Avenue NW,
Washington, DC, 20036,
USA

National (UK)

Africa Centre,
38 King Street,
Covent Garden,
London WC2 8JT

CAFOD,
2 Romero Close,
Stockwell Road,
London SW9

CIIR,
22 Coleman Fields,
London N1

Centre for Global Education,
University of York,
Heslington,
York Y01 5DD

Centre for World Development Education,
Regent's College,
Inner Circle,
Regent's Park,
London NW1 4NS

Christian Aid,
PO Box 100,
London SE1 7RT

Commonwealth Institute,
Kensington High Street,
London W8 6NQ

Commonwealth Institute of Scotland,
8 Rutland Square,
Edinburgh EH1 2AS

Friends of the Earth,
26–28 Underwood Street,
London N1 7JH

Latin America Bureau,
1 Amwell Street,
London EC1R 1UL

London Food Commission,
88 Old Street,
London EC1

Minority Rights Group,
29 Craven Street,
London WC1H 5NT

National Association of Development
Education Centres (NADEC),
6 Endsleigh Street,
London WC1H 0DX

Oxfam (Head Office),
274 Banbury Road,
Oxford OX2 7DZ

Save the Children Fund,
Mary Datchelor House,
17 Grove Lane,
Camberwell,
London SE5 8RD

Third World First,
232 Cowley Road,
Oxford OX4 1UH

VSO,
317 Putney Bridge Road,
London SW15 2PN

War on Want,
37–39 Guilford Street
London SE1 0ES

Women's Information and Resource Centre,
173 Archway Road,
London N6 5BL

World Development Movement,
26 Bedford Chambers,
Covent Garden,
London WC2E 8HA

Index

Tables in bold

accumulation, 142, 156
flexible, 159
primitive, 16, 96, 100
private, 49
rural, 57
advanced economy consumers, depend on Third World exports, 24
Africa
Berg Report, 47–8, 151
food crisis and gender conflict, 54–67
over-simplification of women's role, 8
agrarian populism, 69, 71, 74–5, 77
agrarian reform, Mexico, 104–5
agribusiness, 111, 122, 151
sudden termination of operations, 150
Agricultural Adjustment Administration (AAA), 15, 18
agricultural commercial firms, 94
agricultural contracting, *see* contract farming
agricultural extension work, discrimination against women, 63
agricultural innovation, through migrant wage labour, 56–7
agricultural input industries, 169
agricultural land, value in UK, 185–6
agricultural merchants, activities in India, 101–3
agricultural products
impact of industrial substitutes in Third World, 172
to reflect human and environmental costs, 28–30
agricultural research, 164, 169, 175–6
Agricultural Trade Development and Assistance Act (USA 1954), 18
agriculture, 24, 48–9, 91, 160
African, subsistence and cash-cropping sectors, 55
collective, family labour, 27
India
draining surplus from, 96–101
rainfed and dryland, relatively neglected, 85
industrialization of in Britain, 182–3
Mexico
crisis in, 105–6
monopoly capital, 108–9
need to reorganize "ejidal" system, 109
Tanzania
capital in, 114–16
labour in, 116–18
mainstream view of crisis, 112–14
Third World
for export, 21
reshaping by contract farming, 151–2
underdevelopment of food sector, 20–21
agro-industry, declining opportunities in N. hemisphere, 116
Amazonian rainforest destruction, 194
Anti-Corn League, 24–5
anti-statism, 71
Argentina, 169, 176
Arni, North Arcot, 97–8
Asian NICs, 176, 177
Assam, 88

bananas, 118, 153
Bangladesh
1974 famine, 35–6
donor influences, 39–40
food aid conditionality determining food policy, 34–5
Khulna and Medja Para, 125–32
need for more opportunities for slum women, 133
rural
collective responses to poverty, 145–7
poverty enumeration in, 134–6
purdah, patriarchy, 136–8
renegotiating gender divisions, 141–5
women's survival strategies, 138–41
rural ration system reformed, 41
US food aid, 7
denial of, 34
negotiations for steady supply, 37–42

used to influence government policies, 32–42
US leverage, process of, 36–7
Bangladeshi, effect of Pakistani onslaught, 128–9
Barcelona, contracting to women outworkers, 159–60
barter, 18
basic goods, scarcity of, 119
beef, 181, 182
beefburgers/hamburgers, 28, 181, 194
Belgium, use of fertilizers, 188
Berg Report, World Bank, 47–8, 151
bioprocessing, 170
Biorevolution, 169–70, 178
biotechnology, 11, 51
corporate dominance in, 165–6
decline of Third World exports, 170–2
and the future of Third World development, 178–9
importance of research priorities, 167
for the less favoured, 170
potentially neutral, 167
predominance of First World-oriented research, 167–70
relatively inexpensive, 178
and the Third World: dependence and disadvantage, 163–77
to improve Third World health care and food production, 167–8
biotechnology applications, perverse timing of in Third World, 175–6
biotechnology feedstocks, formation of world markets in, 176–7
biotechnology industry, global, future, 177
biotechnology product imports, 172
Birbhum District, West Bengal, 94
agricultural rents and profits, importance of, 96
dependence on consumption loans, 99–100
grain production, 98
relative technical backwardness, 102
rice-milling magnates and petty traders, 95
role of commercial capital in class formation, 100–101
state-initiated land reform, 102
black market, 117–18, 127, 131–2
Boserup, E., *Women's Role in Economic Development*, 55
Botswana, 182
Brazil, 165, 169, 182
cocoa, 171–2
Britain
diet in, **189**, 190, 191
food policies in, 12
high food costs, 190–91
industrialized food production, 182–3
British Commonwealth Development Corporation (CDC), contracting, 151
British Nutrition Foundation, 191

caciques (chiefs), Mexico, 109
Cameroon, 158, 171–2
capital, 71, 114–16, 149
commercial, 93, 96, 99
roles of, 94, 100–101
mercantile, 91, 93–4
mediating role, 92–3
capitalism, 9–10
exploiting household labour, 160–61
imposing divisions, 7
international, 11, 25–6
as a mode of production, 6
national, India, 97–8
peasants in, 72–5
people's capitalism, 74, 76
political economy of, 3
variations in farming, 76
capitalist countries, advanced, change in farming and diet, 26
capitalist growers, production contracted to, 153
capitalist investment, India, 81
capitalist modernization, in Punjab, 89
capitalist powers, control of policy making and state expenditure, 33–4
Cardenas, Cuauhtemoc, 110
Cardenas, President Lazaro, 104
cash crops, 55
cash income, "off-the-books", Tanzanian women, 117, 121
cashew nuts, 112, 116
casual labour, 116, 117, 118, 121
cell-culture, 170
cereal grains, 80, 169
narrowing of traditional diversified genetic base, 84–5
children, of poor women, 145
class
and gender, 5–6
in markets, 49–50
class differentiation, of the peasantry, 72–3
class formation, 72–3, 93, 100–101
class relations, international, and the international food regime, 24
cocoa, 150, 151, 153
and protein engineering, 171–2
cocoa butter, 171
coffee, Tanzania, 112, 116
Coimbatore District, Tamil Nadu, 94
commercial firm classifications, 95–6
cotton, 98, 99

role of commercial capital in class formation, 100
starting capital for agricultural firms and agro-industry, 96–7
tobacco, pre-harvest loans, 99
traders, controlling trade, assets and storage, 95
collectivization, 27
Colombia, 20, 21, 158, 169
colonial states, 70
commercial farming, 57
commercial firms, misleading classifications, 95
commercial sales, replacing aid, 22–3
commoditization, 8, 74–5, 116–17
effects for women, 56–7
Commodity Credit Corporation (CCC), 15
commodity flows, 94
Common Agricultural Policy, EEC, 15
common income, concept of lacking in Africa, 61–2
community, conceptions of, 69–70
"concessional sales", 18
conjugal dependence, 64
conjugal unit, 138–9
consumer awareness, food issues, 193–4
consumer power, 12
consumer subsidies, removal of, Tanzania, 122
consumption loans, 99–100
contract farming, 11, 52, 149–61
capital and the labour process: logic of the contract, 154–5
contracts, capitalism and unfree labour, 160–61
and global Fordism, factories in the fields, 159–60
on a global scale, 150–52
patterns of in Third World, 152–4
social organization of, 152–4
struggles with the contract, 157–9
contract, the, 155, 159
contracting
exploiting peasant labour market, 154–5
internationalization of, 151–2
contracts, 154–5, 157
abrograted by peasants, 158
arbitrary suspension of, 157–8
and tactical resistance, 158–9
Corn Laws, repeal of, 182, 183–4
corporate dominance, in biotechnology, 165–6
Costa Rica, 153
cotton, 116, 171
in Coimbatore, India, 98, 99
Mexico, 106
Tanzania, 112
counterpart funds, used locally, 18–19
craft products, putting-out basis, 140
credit, 22, 80, 98, 99, 100, 158
crop failure, 130
Cuba, 182

death, and food, 181
debt, among peasant outgrowers, 158
decolonization, 15
deforestation, 85
dependence, creation of, 64–5
dependence trade-offs, 174
deterrence strategy, 183
development planning and projects, role of, 62–4
diet, 21, 181
British, inadequacies of, 184–91
diet/land relationship, 193–4
dietary change, 193–4
disasters, 44
diseases, diet-related, 190
diversification, commercial, 96
divorce, 142–3, 144
domestic servants, 126–7, 128
donor influence, and government policy, 32
drug prices, 174
dualism
in African agriculture, 55
of African women's economic role, 60–61
in agrarian populism, 71
dumping, 14, 21, 25

economic growth, World Bank expectations of, 45
economic instability, Bangladesh, 34, 35
economic/political relations, challenges to, 102–3
Ecuador, cocoa, 171–2
edible wild plant gathering, 139–40
education, collective, and NGOs, 146–7
ejidal system, Mexico, possibilities for reorganization of, 109
ejido land, Mexico, 107
Elson, Diane, on market socialism, 50
empowerment, in Bangladesh, 146–7
Engel effect, India, 84
Ethiopia, Wollo famine, 44
Europe
conjugal dependence increased, 64–5
contract farming, 149
policy of self-sufficiency, 14–15
European Community, seeking to influence food policy, 40
exchange, relations of, 73–4
export crops, 122
exchanged other than through official markets, Tanzania, 112

factory production, eliminated by contracting, 159
families, extended, 137
family
conceptions of, 69
divided, 65–7
family farms
competitive specialization, 27–8
and contract farming, 149
famine, 43–4
Bangladesh, 34–6
sub-Saharan Africa, 55
famine foods, 139–40
farmers
dependent on exports, 24
and labourers, relation to markets, 51
farming, 76
feminization of in Africa, 56
India, investment in new techniques, 86
women's role in, 62
farming contracts, 149
fat, consumption of, 190
fat/protein flows, from Third World, 181, 182
feedstocks
biomass, from low-income countries, 177
plant-derived, 176–7
female dependence, and cohesive family unit, 138–9
female impoverishment, images of, 10
female labour, 116, 117, 118–19, 123
female migration, 145
female-headed households, 57, 58, 65, 66, 144
amongst the poorest, 145–6
Bangladeshi slums, nutritional crisis in, 127
little concern with propriety and purdah, 142–3, 144–5
fertilizers, 98, 184
consumption in UK, 186
removal of subsidies, Tanzania, 122
financial surplus, flows from less to more developed regions, 98
food
in Britain
adulteration of, 189–90
depolitization of, 183, 191–2
real costs of, 190–91
further resources on
books/journals/periodicals, 196–8
useful organizations, 199–202
imported, hidden costs of, 28
mass-produced, 26
political economy of, 2–7
food access, 3, 125
maintained by women, 135–6
food aid
American to Bangladesh, 7, 32–42
assured political stability, 40–41
how it worked, 17–19
integrated Third World into capitalist world economy, 16
and the international food regime, 16–17
and the Marshall Plan, 16–17
multilateral, 23
other donors using leverage, 34
participation of other countries, 22
political/strategic aspects of, 23
undermined local agriculture, 17
food aid conditionality, 33–4, 40
food biotechnology, 166
food capital, in Britain, 191
food consumption, rural India, 84
food crisis, in Africa, 8
food crops, 55, 62
Food First, Lappé and Collins, 1, 48
food grains
China, availability of, 83
India, 81, 82
"self-sufficiency", 83–4
open market sales, 38
food imports, cheap, 21, 182–3
food markets, 47, 49, 50–2
Food Planning and Monitoring Unit (FPMU), 39
food poisoning, 193
food policies, in Britain, 12, 183–4, 185–6, 191–4
food policy management, a new framework, 40
food politics, 13, 27, 192–3
food production
British, 184–5
cost efficiency of, 188
dependency on fossil fuels, 186, **187**
energy inefficiency of, 186
fertilizer use, 188
level of mechanization, 188
packaging and processing of, 188
subsidies, 185–6
resistance to changing terms of, 6
shift to factories, 172
food question, 2, 3, 9
food security
crisis in Africa, 55
local, affected by export contracting, 156–7
food supplies, transnational relations of, 24–5
food systems, long-term impact of biotechnology, 172
food-for-work programmes, 87

foreign exchange, and TNCs, 122
fragrances, 170
fruit/vegetable growing, rural Bangladesh, 140

Gambia, The, irrigated rice, 63–4
gender, concept of, 5
gender conflict
  political consequences, 58
  response to economic stress and poverty, 54
  rural men and women, 65–6
  underlying content, 66–7
gender divisions, renegotiation of, rural Bangladesh, 141–5
gender relations, 3, 4–6, 72
  Africa, 8, 54, 57–9
Germany, Agrarian League, 25
Ghana, cocoa, 171–2
Global 2000 Report, USA, 188
global stagnation, effects on Third World, 164–5
grain markets, in India, 9
grains
  bid away from Third World countries, 177
  cheap, imported into Britain, 182
Great Depression/World War II, collapse of early international food regime, 14–15
Greater London Council, 191–2
Green Revolution, 9, 26
Green Revolution, India, 80–90, 91
  increasing national divisions: Punjab and Assam, 87–90
  national self-sufficiency combined with mass poverty, 82–5
  regional and class concentration of growth, 85–7
  uneven expansion of, 80–82
Guatemala, contract suspension, 157–8

Haiti, 34
Hasna, an illegal trader, Bangladesh, 130–32
health care, and biotechnology, 178
Honduras, bananas, 150
horticulture
  contract volatility, 156
  industry export-oriented, 151–2
household labour, exploited, 160–61
households
  conflict over resource control, 61–2
  conjugally-based, only in planners' models, 63–4
  family-based, 137
  members struggle over rights and obligations, 58
  peasant, wives responsible for self-provisioning, 56
  rural, unable to meet consumption needs, 57
  see also female-headed households
How the Other Half Dies, S. George, 1
hybrid crops, a biological patent, 169

illness, Bangladeshi slums, 129–30, 131
IMF, 3, 25, 33, 40, 151
immigrants, in Britain, 183
imperialism, 32–3, 41–2
imports, effects of low prices, 28
income, variable among poor, 127
India, 9, 165
  agricultural merchants and agricultural change, 91–103
  food stocks, importance of government activities, 86–7
  Green Revolution, some consequences, 80–90
  national self-sufficiency combined with mass poverty, 82–5
  North India, total food output rose steadily, 85
  problems of a union of nationalities, 87
  process of profit-inflation in the economy, 84
  region of stagnating production, 86
  regional and class concentration of growth, 85–7
  self-sufficiency in food grains, 83
  West India, food output declined, 85
  widespread unemployment and poverty, 83
Indian saris, black market trade in, Bangladeshi slums, 127, 128
India's Food Crisis, Ford Foundation, 80
industrial deconcentration, 159–60
industrial development, and foreign aid, 16
industrial substitutionism, 170
industrial tissue culture, 170
industrialization, through cheap food policies, 21
information technologies, new, development of, 164
inputs, subsidized, 184
Intensive Agricultural Development Programme (IADP) (India), 80
International Agricultural Research Centres (IARCs), 164
International Finance Corporation, pioneering contracting, 151
international food markets, 51
international food regime, 13–14, 16–17
  alternatives to, 26–30
  background to, 14–16

end of, 21–4
International Monetary Fund, *see* IMF
International Wheat Agreement, 17
intervention, and markets, 4
investment portfolios, big trading companies, 96
irradiation, 193
irrigation
needed for new dwarf cereals, 84–5
tubewell-pumpset, 85
irrigation pumpsets, 98

Jamaica, 34
Japan, 149, 177, 188

Kabelele, Mary, 118
Kenya, 160
and BAT, 156
extensive production contracting, 152
Mumias sugar scheme, 156
small peasant contracting, Njoro Canners, 153–4
smallholder schemes, 156
study on nutrition, 157
tea production, 151
vegetable contractors and "leakage", 158
Khulna and Medja Para, Bangladesh, 125–32
Hasna, an illegal trader, 130–32
Mina, a Hindu widow, 128–30
struggles for survival of the poor, 126–8
Korean War, 17, 18
kulak class, Coimbatore, 99, 100

labour, 6–7, 11, 116
casual, 57, 116, 117, 118, 121
female, 116, 117, 118–119, 123
rural, North India, 81
sexual division of, 55
in rural households, Tanzania, 116–17
through domestic and kinship arrangements, 59–60
Tanzania
in agriculture, 116–18
increasingly female, 112–13
protests against sexual division of, 118–19
unfree, 155, 160–61
see also wage labour
labour markets, 51–2, 117
labour migration, 56–7
land, needs of animal protein, 181
landlord-capitalist growth of power, effects of, Punjab, 89–90
landlords, Bangladeshi slums, 126
Law of Agricultural Development (Mexico), 107

Liberia, US economists in government, 33
"like-minded group", alternative agenda, 40
livestock, share-rearing, 140, 143–4
living standards
improved by biotechnology, 178
need to raise in Third World, 29
loan forgiveness, 37, 39
local currency funds, 39
localism, in food, 29
London Food Commission, 192

maize, 37–8, 112, **187**
hybrid, 169
Malawi Tea Authority, 156
Malaysia, 151, 171–2
Mali, power of food donors, 33–4
manufacturing bourgeoisie, 89
market liberalization, 74–5
market makers, 50
market socialism, 50
market specification contracts, 154
markets, 3–4, 8, 43
abstract and real, 46–52
agricultural, 96
and agricultural change, 91
can be ignored, 92
competitive and allocatively efficient, 92
development or abrogation, 51–2
in feedstock materials, 177
for food and agro-industrial crops, 91
inefficient and monopolistic, 92
and non-market relations, 44
power and control over terms of operation, 51
real and real needs, 48–52
as sets of social relations, 50
structured by state action, 50
superior efficiency assumed, 45
views about, 92–4
World Bank meanings, 46–7
marriage, rural Africa, 58–9, 61–2, 65
Marshall Aid/Plan, 15, 16–17
meat products, cheap, 24
mechanization, 188
men, not accepting reponsibilities towards dependants, 142–3, 144
Mexico, 165, 169
agrarian history of, 9–10
agrarian and political crises, 104–10
calls for change in system of government, 110
capitalist development, 105
contract farming, 150
contracting to women outworkers, 159–60
land distribution, 104, 106
Nestlé

Chiapas, milk suppliers, 155
relocation of operations, 158
political crisis, 108–10
sharecropping, 160
socio-political change, 110
middle peasantry, 72, 151, 156
migrant labour, 89, 116
migration, Mexican, to USA, 110
Mina, Hindu widow, Bangladesh, 128–30
Ministry of Agriculture, and food capital, 191
minorities, reaction against, 87
MNCs (Multinational Corporations), *see* TNCs
molecular and cell biology, 172–3
moneylending , 97, 98–9
moneylending capital, 93
Mozambique, poor market regulation, 49

National Confederation of Peasants (CNC), Mexico, 109
NATO, 17
natural resource degradation, 25–6
neo-classical populism, 75
The Netherlands, use of fertilizers, 188
NGOs, and empowerment, 146–7
Nicaragua, 182
Nigeria, cocoa, 171–2
nitrogen fixation, 166
non-wage labour, 161
North Arcot District, Tamil Nadu
commercial firms lending money for production and concumption, 98–9
commercial profits, 97
rice wholesalers, 94
role of commercial capital in class formation, 100
starting capital, 97
state and the grain trade, 102
transformed techniques of production, 98
North-South disparities, scientific and technological, 163–4, 178
nucleus-estate schemes, contract farming, 153, 154
nutrition, lower for females, rural Bangladesh, 138

oils, low-priced, conversion of, 171
Ordaz, Diaz, government (Mexico), 106
outgrower schemes, 151, 154, 156
outsourcing, multiple, 159
overconsumption, 1
overproduction, 1, 106

Pakistan, 34
palm oil, 150, 151
parallel markets, 49
patriarchal authority, Tanzanian women speak against, 120–21
patriarchal bargain, rural Bangladesh, 137–8, 144
patriarchal relations, oppressive, Tanzania, 117
patronage, Bangladeshi slums, 129
peasant farmers, and TNCs, 3
peasant/worker incomes, pressure on, 123–4
peasantry
class differentiation in, 72–3
views of, 8–9
peasants
adaptation to modern life, 70
in capitalism, 72–5
exploited through terms of trade, 74
independent, producing for home or export, 150
India, compelled to borrow, 99
Mexican
"below subsistence", 108
dispossession of affecting society, 110
felt tied to state, 104
proletarianization of, 108–9, 110
role in political system, 108–9
ruined, 105
oppression of, 76
petty-commodity producing, 100
and socialism, 75–8
and socialist co-operation?, 76–7
view of, 69
working as piece workers, 160
see also poor peasants
peasants (virtuous), and states (vicious), 69–71
abandonment of, 77–8
Peru, 34, 158
pesticides, 80, 98
petty commodity production, 72, 139
petty trade, 124, 126, 127
petty traders, 95
pharmaceuticals and health care, Third World and OECD, 168
pharmceutical industry, comparative analysis (Third World), 173–4
Philippines, 151, 169
pioneer farmers, independent, producing for home or export, 150
planning, 49
plant molecular biology, research and development, 165–6
plant protection, non-chemical means, 168
plant-derived chemicals and materials, and biotechnology, 170
plant-derived materials, future use of, 177

plantation/estate production,
decomposition of and contracting, 150
plantations, "runaway", 111
political intervention, 92
polygyny, increased, 61
poor
rely on own labour power, 139
see also rural poor
poor peasants, 73, 156
poor women, seek fall-backs, 144
population growth, India, 86
populism and efficiency, Mexico, 107–8
Portillo, Lopez government (Mexico), 107
poultry, 140, 143–4
poverty
Bangladesh, 126, 145–7
feminization of, 5–6, 125
linked to desertion, divorce and
female-headed households, 142–3, 144
a source of domestic conflict, 66–7
poverty enumeration, Bangladesh, 134–6
Poverty and Famines, A.K. Sen, 43–4, 45
Poverty and Hunger, World Bank, 44,
45–6, 51
poverty measurement, 134
power, 33
in markets, 49–50
price controls and rationing, Tanzania, 113
price guarantees, 190
price support, agricultural products, UK,
185
primary production, 177
private sector, 45–6, 47
biotechnology research, 164, 165, 167
processed food, consumption in UK, 188
processing, economies of scale, 153
producers, impoverished, 25
production management contracts, 154
profit repatriation, 173
proletarianization, 86, 108–9, 110, 116–17,
123
proletariat, rural, commercial credit
for subsistence, 100
protectionism, in UK, 186, 188, 190–91
Public Law 480 (USA)
Title I, provisions of, 18–19
Title III food aid, 36–7, 39
pulse production, India, 81
Punjab, 86, 88–90
purdah
at the ideological level, 137
constraints of, 136–7
not overridden by poverty and
landlessness, 138
and women's survival strategies, 139–41
putting-out system, textiles, 160

radicalism, right-wing, in Assam, 88
recombinant DNA production, 170–71
recycling, 188
reformed sector, Mexico, 107
regulatory interventions, benefits from
misimplementation of, 102
religious fundamentalism, 87
remittances, to Punjab, 89
resource ownership, 50–51
resource-providing contracts, 154
rice
Gambia, 63–4
India, 81, 82
rural Bangladesh
crumbling of gender rules, 141–2
man/woman labour division, 136–7
Tanzania, 114
rice gleaning, 139
rice processing, post-harvest, 140
rich peasants, 73
rubber, 151
rural depopulation, 26
rural development projects, 75
rural economy, great changes in, 57
rural employment generation, 122
rural poor, 151, 152
India, 84, 87
rural renewal, through urban
consumptionism, 181–94
rural wage bill, India, 84
rural women, 141
migration of, Bangladesh, 125
in Tanzania, 118–22
rural-urban differentials, Arni, North
Arcot, 97–8

Scott, Chris, neo-classical populism, 75
secession, mooted in Punjab, 87, 88–90
security, 52
self-exploitation, 72
self-sufficiency, 15, 83
Senegal, 150, 154, 182
services
importance of to rural women, 113
"user fees" for, 122
settler territories, cheap food production,
183
share-rearing, 140, 143–4
gender implications of animal choice, 144
sisal, Tanzania, 112, 114, 116, 121
Sistema Alimentario Mexicano (SAM),
107–8
slums, Bangladesh, 125–32
small firms, Birbhum
dependence on large firms, 95
illegal, 100–101
small peasant contracting, 153–4

small-scale farmers, and biotechnology, 170
smallholder development/schemes, 151
  growers differentiated, 156
social convention, and women's survival stategies, 138–9
social differentiation, Tanzanian rural areas, 118
social organization, of contract farming, 152–4
social relations
  conventionally different financial responsibilities, 66–7
  of food, 2–3
social security, preventing famine in the West, 44
social services, views of Tanzanian women, 120
social transformation, 77
socialism
  bureaucratic, 76, 77
  and peasants, 75–8
socialist alternative for agriculture, 26–30
socialist co-operation, 77
socialist countries, subsidized imports, 25
socialist governments, re-examination of role of markets in the economy, 49
socialists, 4, 71
  and markets, 3–4, 8, 49–50
socio-economic differentiation, 178
sorghum, 150, 158
  hybrid, 169
"speak bitterness" sessions, Tanzania, 118–22
spices, 170
Sri Lanka, 34
state corporations, refusal to pay adequate wages, 121
state economic activity, 70–71
state, the
  ambivalent towards commercial sector, India, 102–3
  power of, 159
  and unfree labour, 160–61
state trading, 92
states, viciousness of, 70–71
stereotyping, 62, 134–5
  of African rural women, 56
  of peasants, 70
  of poor women, 135
  of women, rural Bangladesh, 137
strategic raw materials, 18
strawberry imperialism, 24
structural adjustment, 33, 122–4
Structural Adjustment Programmes (SAP), 114, 122
subsistence sector, African agriculture, 55–6
sugar, 150, 153, 170–71, 189–90
  Kenya, 157
  Mexico, 106
  Tanzania, 112, 114, 121
supermarkets, adjusting easily to food issues, 193
surplus appropriation, 93–4, 99–100
  secondary, 94, 100
surpluses, 96, 193
survival strategies, 135
  male, abdication of family responsibilities, 142–3
  of women, 6, 10, 138–41
Swaziland, 154, 155
sweetners, 170

Tanzania
  cost of living, 118, 122
  cropping, **115**
  deregulation of import/export trade, 122
  devaluation, 122
  food staple sales, local and parallel markets, 112
  government blamed for inadequacy of village services, 119–20
  investment in most productive regions and areas, 122
  mainstream view of agricultural crisis, 112–14
  non-farm commodity activities, 117–18
  production, large-scale, 114
  rural women speak, 118–22
  structural adjustment, 111–24
TB, in Bangladeshi slums, 129–30
tea, 114, 150, 153
  Kenya, 151
  Malawi, 156
  Tanzania, 112, 116, 121
technological inferiority, and biotechnology, 163–5
technology transfer, 151, 173, 174
tenant evictions, India, 81
tenurial reforms, Birbhum, 100
terrorism, Punjab, 88, 90
Thailand, 151
Thatcher, Margaret, 192
Third World
  agro-food complexes in, 149–61
  biotechnology
    impacts of, 175–6
    potential long-term benefits of, 172–3
    significance of private-sector dominance in, 166
  biotechnology and agricultural development in, 163–79
  capitalist strategies of domestic food production, 25
  cheap food policies, 19

class structure of trade, 51
contract farming, 11, 149–61
patterns of, 152–4
dependence on subsidized grain imports, 16
development, 70
dilemmas of capitalist development, 18
export decline, and biotechnology, 170–72
a field of competition, 22
food import dependence of, 14
hunger in, 1
importance of state policy, 51
and inequality, 6
limits to market expansion, 25
new export markets created, 14
patterns of superiority-inferiority within, 165
socialist governments and markets, 49
technological inferiority, new pattern, 164–5
upward and downward mobility of nations, 176
TNCs, 165
benefitting from Structural Adjustment Programmes, 122
contracts favour those with capital, 155
investment in biotechnology, 173–4
and the peasant farmer, 3
pharmaceutical, involvement in Third World, 174
profitability eroded by crop "leakage", 158–9
refusal to pay adequate wages, 121
in Tanzania, 112, 114, 116
tobacco
in Coimbatore, India, 99
Kenya, 157
Tanzania, 112
trade
agricultural, structures of, 94–6
class structure in Third World, 51
international, change in pattern of, 16
opportunities for women, 57
role of aid in restructuring, 19–20
separated from aid, 22
trade embargoes, 16, 17
trade wars, 22
trade-offs, gender-constrained, 143
traders, class power of, 49
trading profits, Coimbatore, 97
transnational corporations, *see* TNCs

UN Food and Agricultural Organization (FAO), 164
underdevelopment, and proletarianization, 19–21
unemployment, increased by biotechnology, 170–71
urban consumptionism, a route to rural renewal, 181–94
urban food markets, supplied by cheap imports, 21
urban growth, Bangladesh, 125, 133
urban migrants, female, 58
urban migration, 123, 125, 126, 133
urban population, growth of, 21
urbanization, early, Britain, 183
USA, 32
contract farming, 149
domestic farm policy, 13–14
dominant export country, 16
financing international trade, 15
food aid conditionality, 41–2
food aid negotiations, 37–8
food surpluses, 7
influencing recipient countries through food aid, 33–4
obtained local currencies through Title I food aid, 18, 19
political strength of farmers, 15, 23–4
promotes capitalism as aim of aid-giving, 45
share of world food aid falling, 23
wheat, 7, 13, 15, 17, 21–2
USAID, 34, 37, 40, 41, 151
USSR, a wheat importer, 16

vertical co-ordination, *see* contract farming, 149
village resources, control of, 119

W. Germany, use of fertilizers, 188
wage labour, 65, 126, 139
decline in alternatives to, 123–4
wage work, casualization of, 7
water conservation systems, smaller, decline in, 85
water development projects, 39
water-table, lowering of, 85
wheat, 7
black market in Bangladesh, 131
exports, 13–14
India, 81
in the Punjab, 88
rise in world price, 22
surplus stocks, effect of, 13–14
Tanzania, 112, 114
USA, 7, 13, 15, 17, 21–2
a vehicle of proletarianization, 21
in world trade, 20
women
Africa

economic role in the farm family, 59–62
and food farming, 8
lack of capital for independent farming, 60–61
negative impact of development, 56
rural accumulation patterns, 57
Bangladesh
employment opportunities restricted in slums, 126–7
subordination of, 146
well-being of, and patriarchal bargain, 137–8
covert earnings, 143
dependence of, 123
discriminated against in agricultural extension work, 63
effects of commoditization, 56–7
as family labour, 65
images of, 5
important in sub-Saharan food production, 55
lack income distribution control, 119
poor, not passive, 135–6
in public rural work projects, 142
and rural development projects, 62
rural to rural migration, 57
seen as a vulnerable group, 135
in small peasant contracting, 154
struggle for access to food, 10
and a subsistence sector?, 55–6
Tanzania
pushed into casualized agricultural labour, 10
refusing estate work, 121
undermining male authority, 121
try to gain autonomy within domestic hierarchy, 143
see also rural women
"Women in Development" lobby, 55
Women in Development (WID) programmes, 10, 123
women farmers, dilemmas of, 65
women's co-operative shops, 113
women's groups, challenges of, 147
women's income, use of, rural Bangladesh, 142
women's labour, contribution significant, low-income households, 139
Women's Role in Economic Development, Ester Boserup, 55
women's survival strategies, 6, 138–41
workers, dispersed, technical and social integration of, 159–60
World Bank, 3, 33, 40, 52, 111, 123
Berg Report, 47–8, 151
integration with IMF lending, 45
meaning of markets, 46–7
Poverty and Hunger, 44, 51
on role of private sector, 45–6
in Tanzania, 10, 112, 114
World Development Report, 44–5
world economic system, 1
world food economy, possible restructuring of, 172
World Food Programme of the UN, 23
world trade, and American aid, 20